INTRODUCTION TO MANUFACTURING PROCESSES AND MATERIALS

MANUFACTURING ENGINEERING AND MATERIALS PROCESSING
A Series of Reference Books and Textbooks

FOUNDING EDITOR

Geoffrey Boothroyd
University of Rhode Island
Kingston, Rhode Island

INTRODUCTION TO MANUFACTURING PROCESSES AND MATERIALS

ROBERT C. CREESE
West Virginia University
Morgantown, West Virginia

CRC Press
Taylor & Francis Group
Boca Raton London New York

CRC Press is an imprint of the
Taylor & Francis Group, an **informa** business

CRC Press
Taylor & Francis Group
6000 Broken Sound Parkway NW, Suite 300
Boca Raton, FL 33487-2742

First issued in paperback 2019

ISBN-13: 978-0-8247-9914-4 (hbk)
ISBN-13: 978-0-367-39989-4 (pbk)

Library of Congress Cataloging-in-Publication Data

Catalog record is available from the Library of Congress

Visit the Taylor & Francis Web site at
http://www.taylorandfrancis.com

and the CRC Press Web site at
http://www.crcpress.com

Preface

The traditional approach to manufacturing processes is that of a "seed catalog," in which each process is presented in a separate section. This approach makes it difficult to compare processes or to get a general overview of manufacturing. This book takes a new approach, integrating manufacturing management, materials, manufacturing processes, and design considerations. This presentation also places an emphasis on problems, because engineering students are problem solvers and enjoy the solution of problems.

Manufacturing is multidisciplinary—it involves the product-design and strength-of-material aspects of mechanical engineering; the fundamentals-of-material-structure, solidification, and process-design concepts of materials processing and engineering; and the cost-evaluation, fundamentals-of-manufacturing-processing, quality-control, and total-quality-management aspects of industrial engineering. Other topics, such as tool engineering and failure analysis, have been taught in more than one of the aforementioned areas. Because manufacturing involves such a broad spectrum of knowledge, it requires integrated teamwork among members of the various engineering disciplines. Teamwork—involving design, materials selection, process selection, marketing, purchasing, production, quality, and other manufacturing functions—is necessary to develop the "best" product for the consumer.

A new systems approach to process selection is presented to indicate the importance of the integration of design, materials, and product manufacture. The relationship of material properties to design is described, along with approaches to material selection considering cost. The discussions in Chapters 2 through 5 represent a review of basic materials science and engineering material properties, and students who have had previous courses in materials may need only a quick review of these subjects; however, some items presented in this text are often omitted in traditional materials and material property courses. The information in Chapter 6 is very important and has not been included in most materials science or engineering design courses.

This is an introductory book, and it focuses on traditional metal manufacturing for two primary reasons:

1. Metals are used more than any other material, and their material properties, such as strength, density, and cost, are better known that those for the other commonly used structural materials, such as polymers, ceramics, and wood. The design relationships are better developed for metals, and thus it is easier to understand the integration of design, materials, process, and management.

2. This is an introductory book, and the relationships for design and manufacturing are quite different from those used in electronics manufacture or for complex composite structures. These more specialized topics must be covered in more specialized books. However, the design relationships that use volume fractions for microstructures and property determinations are the same as those used to determine the properties of composite mixtures.

The students in the IMSE 202 classes at West Virginia University were very helpful with their critical comments concerning most of the material presented. Dr. Sheikh Burhanuddin, a former graduate student in the Industrial Engineering Department, did most of the work on preparing the materials for Chapter 10 on polymer processing and Chapter 16 on tool design for manufacturing. Most of the figures in this book were created by Dr. Mansoor Nainy Nejad, a former graduate student in the Industrial and Management Systems Engineering Department.

During the writing of this book, the World Wide Web has become a major information source, and considerable information on manufacturing and materials is now available on the Web. In addition, the author has been working with Dr. Poul H. K. Hansen at Aalborg University in Denmark on the development of a "virtual textbook" on manufacturing processes. Mr. Naveen Rammohan selected most of the Internet references utilized in this book. The Internet references for the Aalborg University Process Database being developed by Dr. Poul H. K. Hansen and the IMSE 304 Process Database at West Virginia University are, respectively:

> http://www.iprod.auc.dk/procesdb/index.htm
> http://www.cemr.wvu.edu/~imse304/

Many people who truly believed in the importance of manufacturing have had an indirect influence on this book: Ben Niebel, Alan Draper, and Jerry Goodrich from the Industrial Engineering program at Penn State; Paul DeGarmo during my M.S. program at Berkeley; and Erik Pedersen during his visit to West Virginia University and my visits to Aalborg University in Denmark. Other faculty members at Aalborg University have been helpful, especially Poul Hansen and Sven Hvid Nielsen. The importance of materials was emphasized by George Healy, George Simkovich, and "Doc" Lindsay during my Ph.D. program in metallurgy at Penn State.

There are many new concepts in this book, and the emphasis is on problem solving. The design of the product is only the starting point in manufacturing; for example, for a casting shrinkage allowances must be added, the feeding (riser) and gating systems must also be designed, the pattern must be produced, and the product design may need to be modified to reduce manufacturing costs. There will be errors in this book and they are solely my responsibility and not that of the others who have written or assisted in the writing of various sections. Their help was greatly appreciated.

The writing of this book has taken time from other activities, particularly the "free" time I would have spent with my family members— Natalie, Jennifer, Rob and Denie, and Chal and Carol—and their patience has been appreciated. Finally, it is appropriate to dedicate this effort to our most recent family members, our grandchildren Robby and Samantha, who continually remind us of the wonderful future ahead.

Robert C. Creese

Contents

Contents

I
MANUFACTURING MANAGEMENT AND OVERVIEW

1

The Role of Manufacturing in Global Economics, Manufacturing Aids, and Manufacturing Break-Even Analysis

1.1 INTRODUCTION

Manufacturing, which has been practiced for several thousand years, is, in the broadest sense, the process of converting raw materials into products. The word *manufacturing* is derived from the Latin, *manu factus*, which means "made by hand." Reference to manufacturing is made in Genesis (4:22), when Tubal-Cain is described as a smith who made sharp tools of bronze and iron. This not only indicates the long history of manufacturing, but also indicates the importance of materials in manufacturing which has often been neglected.

Manufacturing engineering is the term widely used in the United States, whereas in Japan and Europe the term *production engineering* is used. Manufacturing has always been a critical element in warfare; but during the Industrial Revolution of the 18th century, manufacturing became a critical element in society by providing consumer goods and mechanizing agriculture to reduce the number of people needed to produce the food supply. During the Industrial Revolution of the 20th century, much of the physical labor and dangerous work is being performed by machines rather than by human effort.

The relationship (1,2) between the gross national product (GNP) per capita and the contribution of manufacturing to the GNP is presented in Fig. 1.1. Countries with high manufacturing contributions to the GNP have higher per-capita incomes than those countries with low contributions of manufacturing to the GNP. In addition, recent events in world politics have indicated the importance of a strong manufacturing base. The rise of Japan and other nations of the Pacific rim has been the result of their strong manufacturing base, and the decline of Eastern Europe and the breakup of the Soviet Union

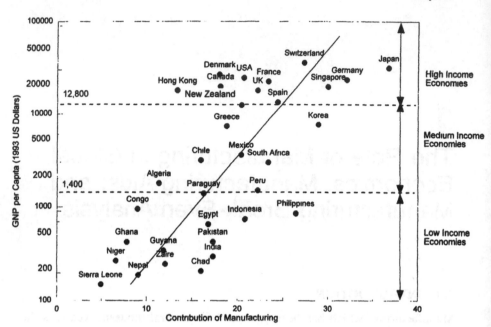

Figure 1.1 GNP per capita versus manufacturing contribution. (Developed from data in Ref. 2.)

have partly been the result of a weak economic base. The low-income nations (less than $1400 per capita) tend to have low manufacturing contributions (less than 15 percent) to their GNP. The high-income nations (greater than $12,800 per capita) tend to have a higher contribution (approximately 25 percent) of manufacturing to their GNP.

The manufacturing base of the United States has declined as a large loss of manufacturing jobs has occurred in the coal, steel, automobile, foundry, electrical machinery, computer, and other manufacturing segments. During the 10 years from 1978 to 1987, the contribution of manufacturing to the GNP of the United States decreased from 25.5 percent to 20.7 percent. Although jobs have been created in the service, health care, and government areas, these jobs have not been as productive to the economy. If the manufacturing capabilities of the United States does not increase, it will no longer be "the economic superpower" of the world. The military power, even the "superpower" status, of the Soviet Union did not prevent the collapse of its government. A strengthened manufacturing base is essential for the United States to maintain its position as a world economic leader.

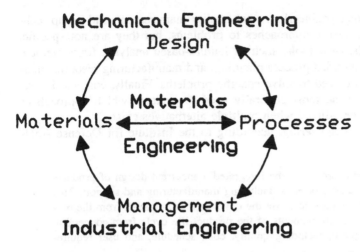

Figure 1.2 Relationships and interactions between the manufacturing functions and engineering disciplines.

The problems of health care, education, crime, and the elderly have increased as the traditional manufacturing base has decreased. Families are no longer able to be supported by a single wage earner, which was common prior to the 1970s.

Manufacturing is not only the making of product; it is the integration of product demand, product design, material selection, manufacturing processing, product assembly, and management to produce a desired product at a competitive price at the desired time. In the past, manufacturing has been considered as a single, separate sequential step in the production of a product; but now it begins as the product is being conceived and lasts until the product is being recycled. Product disposal is becoming more important as environmental concerns become critical.

Figure 1.2 shows the relationships and interactions between manufacturing and engineering disciplines. The figure represents the association between design, materials, processes, and management and their interaction to produce a product. It also depicts the type of engineering disciplines involved and their role in the manufacturing environment.

1.2 MANUFACTURING AIDS

There are many new tools and philosophies to aid in the solutions of the many problems faced in manufacturing. The manufacturing philosophies, such as concurrent engineering, total quality management, and group technology, provide overall approaches to problems, but they are not specific, specialized tools. New tools, such as feature-based analysis, finite-element models, computer-aided process planning, and manufacturing resource planning (MRP-II) are used to solve specific problems. Finally, cost and break-even analysis are the tools generally used to evaluate which approach or solution should be utilized when multiple alternatives exist.

Concurrent engineering, according to the Institute for Defense Analysis (3), is

> a systematic approach to the integrated, concurrent design of products and their related processes, including manufacturing and support. This approach is intended to cause the developers (designers), from the outset, to consider all elements of the product life cycle from conception through disposal, including quality, cost, schedule, and user requirements.

This definition emphasizes the product life cycle, because for military purposes the support and disposal functions are extremely critical. However, for most commercial products, it is difficult to evaluate disposal aspects properly, for we currently live in a "throwaway" society and the customer, rather than the manufacturer, controls product disposal. In the areas of nuclear weapons and hazardous materials, the disposal issue is being addressed in the design stage.

The philosophy behind concurrent engineering is that improvement of the design, production, and support processes are never-ending responsibilities of the entire enterprise. A major difference between concurrent engineering and traditional engineering is in problem methodology. Concurrent engineering applies a multidirectional information approach, resolving design, material, and production problems in an integrated fashion. The design is not complete until the material and manufacturing concerns have been addressed. In the traditional engineering approach, a sequential information approach is used, whereby the design concerns are addressed first, then the material concerns, and finally the manufacturing concerns. The traditional approach often resulted in manufacturing nightmares, frequent costly manufacturing change orders, and high scrap rates, which lead to high costs and low profitability.

For concurrent engineering to be successful, teamwork and good lines of communication are important (10). People from different departments interact over the production life of the product. For example, design engineers and manufacturing engineers must work as a team to anticipate problems and eliminate them before they actually occur. A typical problem of the traditional approach is that design engineers design a part without considering the practicability and economics of manufacturing that part. In a concurrent engineering environment, the design engineer works with the manufacturing engineer to produce a design that includes the manufacturability and the economics as well as the functional purpose.

Group technology (*GT*) is the manufacturing philosophy that identifies and exploits the underlying sameness of parts and manufacturing processes (4). It groups together parts that require similar operations and machines (9). In a manufacturing system, group technology provides for an economical way of producing a part. The majority of products are produced in batches of 75 units or fewer, and this prevents the attainment of the economies of scale of line production. However, with the application of group technology, part families can be created; and with the use of manufacturing cells, the economies of line production can be obtained even though individual products occur in small batches or lots. The use of classification and coding of parts is one of the tools of group technology that permits the classification of parts into part families. Part families permit more rapid throughput and reduce the number of setups, the length of waiting, and the number of movements between operations. This not only reduces product cost but also improves customer service by getting the parts to the customer sooner.

Total quality management (*TQM*) has been defined (5) as a "leadership philosophy, organizational structure, and working environment that fosters and nourishes a personal accountability and responsibility for the quality and a quest for continuous improvement in products, services, and processes." It defines quality based on the needs of the customer. Like concurrent engineering, TQM requires a holistic approach, meaning the entire organization must be involved. Quality can no longer be considered only the Quality Department's responsibility, but everyone's, from top management to the shop floor. The importance of accountability and responsibility is not for the assignment of blame when defects occur, but to determine who has the authority for correcting the causes of the defects and what monitoring is required to prevent their recurrence. With rapidly changing technology and more complex product designs, manufacturing problems will occur, but they must be found, defined, and solved in a rapid manner and prevented from recurring. One of the tools of total quality management is statistical process control (SPC). Statistical process control is the use of statistical techniques to control a process by predicting if the deviation is random variation or if something in the process has changed. The ISO (International Standards Organization) is having an effect worldwide by requiring third party audits of manufacturers to assure that the manufacturing processes are being performed as documented by the manufacturer.

Computer-aided process planning (*CAPP*) has reduced the time for generating process plans. A *process plan* contains the "instructions" for producing a part. It establishes which machining processes and parameters are to be used in order to develop a final product from the raw materials. Process plans have traditionally been made by a process planner using his or her knowledge and experience to determine the appropriate procedures. The arrival of CAPP systems has made the job much easier. Two approaches have been developed, the variant planning approach and the generative planning approach. In the variant approach, a database of process plans is kept on hand; and when a new part is to be produced, the computer searches the database for a similar part and modifies it. It is estimated that this approach can save up to 90 percent of the process planning time (6). The generative process plan approach designs the process plan for the new part automatically and starts from "scratch" rather than via modification of an existing plan. In theory .the generative approach should produce better plans, but generative systems have not been fully developed. For either approach, the integration of the computer-aided design (CAD) drawings is necessary for efficient operation. The goal of CAPP is to be able to generate the process plan automatically from the CAD drawing of the part.

Feature-based design is one of the geometric modeling techniques that has been adapted for manufacturing. The initial emphasis has been upon machining, but it is being adapted to other processes, such as casting. A manufacturing feature describes a geometric entity (solid, surface, line, or point in the component or tool design) that is of interest, considering manufacturability. In a casting, some of the features of interest are: (1) the parting line, which divides the mold into the cope and drag, and (2) depression features, which are used in the design of cores (7). Some of the interfaces and contours for a casting are indicated in Fig. 1.3, which presents the relations between the casting, the core, the pattern, and the mold cope and drag. Thus the influence of a change in the casting surface will result in a feature change that can effect the corresponding features of the core, the pattern, and the mold. One advantage of feature-based design is that the effects of the design changes upon the process can be evaluated much more rapidly. In addition to the product design, other casting design requirements include the gating system design and the riser (feeder) design for the casting.

Activity-based costing (*ABC*) is an accounting approach that assigns costs to activities that are utilized in producing products. This procedure attempts to determine more realistic product costs to assist management in evaluating the costs and profitability of products. In most instances, this procedure has not replaced traditional accounting for tax reporting, but it

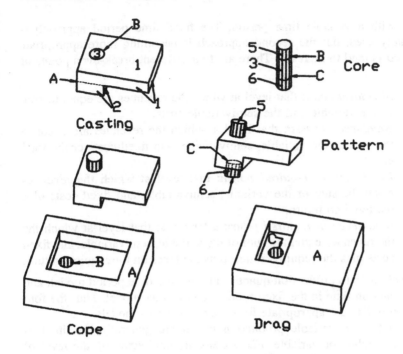

Interfaces | Contours
1 Casting : Cope | (Intersection of Interfaces)
2 Casting : Drag | A. Casting : Cope : Drag
3 Casting : Core | B. Casting : Cope : Core
4 Cope : Drag | C. Casting : Drag : Core
5 Cope : Core | D. Cope : Drag : Core
6 Drag : Core |
(Planes) | (Lines)

Figure 1.3 Interfaces and contours for castings. (Adapted from Ref. 7.)

has given managers a more accurate view of their product costs and the
effect of overhead distribution to products.

1.3 MANUFACTURING BREAK-EVEN ANALYSIS

There are two basic approaches (8) to break-even analysis, that of the fixed
time period with a variable production rate and that of a fixed production

quantity with a variable time period. The fixed time period approach is traditionally used, but the second approach is becoming more appropriate as fixed costs tend to increase. There are four different break-even points of interest:

1. *Shutdown point*: that level at which the revenues are equal to sum of the variable and the semivariable costs.
2. *Break-even at cost*: that level at which the revenues are equal to the sum of the variable, semivariable, and fixed costs or the total costs.
3. *Break-even at required return*: that level at which the revenues equal the sum of the variable, semivariable, and fixed costs plus the required return.
4. *Break-even at required return after taxes*: that level at which the the revenues equal the sum of the variable, semivariable, and fixed costs plus the required return plus the taxes on the required return.

Level refers to production quantity in the fixed time period model, and to the production time in the fixed production quantity model. Thus the four break-even points are appropriate for either of the two models.

Costs for the variable production model are generally classified as fixed, semivariable, or variable. *Fixed* costs are unaffected by the level of production; some typical examples are property taxes, insurance, plant security, and administrative salaries. Production time is a fixed cost. *Variable* costs vary directly with the level of production; typical examples are direct material costs and direct labor costs. *Semivariable* costs generally have a fixed and a variable component; for example, plant maintenance costs require a minimum staff, but as production increases, more maintenance is required and the maintenance costs increase. Required return would commonly be considered a fixed amount, but in some cases it may be more appropriate to consider it a variable or semivariable item.

1.3.1 Variable Production Level with Fixed Time Period

An example problem is presented to illustrate the calculations of the various break-even points. This is illustrated for the planned production period of 100 hours and an anticipated production level of 1000 units. The revenue/cost data of Table 1.1 applies to the example problem.

The *shutdown point* is found by:

revenue = semivariable costs + variable costs

$$10x = 1x + 400 \qquad + 4x$$
$$5x = 400$$
$$x = 80 \text{ units} \tag{1.1}$$

Table 1.1 Cost/Revenue Data for Break-Even Points with Fixed Time Period

Item	$/Unit	Dollars ($)	Decimal
Revenue	10		
Fixed Cost		2000	
Variable Cost	4		
Semivariable Costs	1	400	
Required Return		600	
Tax Rate			0.40

The *break-even at cost* is found by:

revenue = total costs

revenue = semivariable costs + variable costs + fixed costs

$$10x = 1x + 400 \qquad + 4x \qquad + 2000$$
$$5x = 2400$$
$$x = 480 \text{ units} \tag{1.2}$$

The *break-even at required return* is found by:

revenue = total costs   + required return

$$10x = 1x + 400 + 4x + 2000 + 600$$
$$10x = 5x + 2400 \qquad + 600$$
$$5x = 3000$$
$$x = 600 \text{ units} \tag{1.3}$$

The *break-even at required return after taxes* is found by:

revenue = total costs   + required return/(1 − tax rate)

$$10x = 1x + 400 + 4x + 2000 + 600/(1.00 - 0.40)$$
$$10x = 5x + 2400 \qquad + 1000$$
$$5x = 3400$$
$$x = 680 \text{ units} \tag{1.4}$$

At the anticipated production level of 1000 units, the revenues and costs to be expected are as indicated in Table 1.2.

The four break-even points are illustrated in Fig. 1.4, which is a plot of the costs and revenues as a function of production quantity. Since the production level of 1000 units is greater than the break-even point of required return after taxes (680 units), the profit of $1560 is greater than the

Table 1.2 Revenues, Costs, Taxes, and Profits for a Production Level of
1000 Units

Revenues		$10,000
Costs		
Fixed	$2000	
Variable	4000	
<u>Semivariable</u>	<u>1400</u>	
Total Costs	7400	<u>7,400</u>
Gross Profits		2,600
Taxes (@ 40%)		<u>1,040</u>
Net Profits		1,560

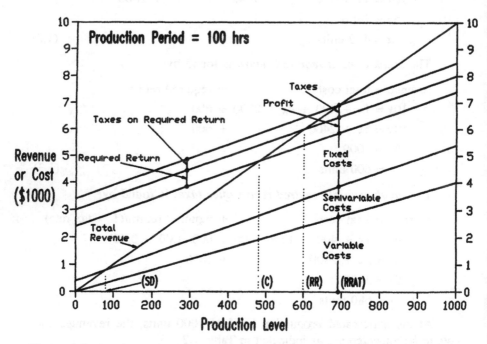

Figure 1.4 Break-even points as a function of production level for a fixed production period.

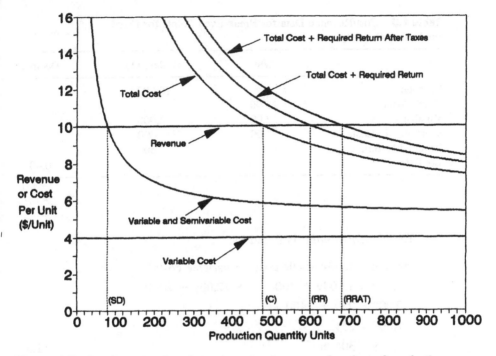

Figure 1.5 Break-even points for unit cost values as a function of production level.

$600 amount of the required return after taxes. Figure 1.5 is a plot of the unit cost values versus the production level. The effect of increasing the revenues on the break-even points can easily be shown with this plot. The break-even points are the same on both Figs. 1.4 and 1.5; the difference is whether total costs or unit costs are used.

1.3.2 Variable Production Time and Fixed Production Level

The traditional fixed costs are fixed over a time period (100 hr), so if the time period is variable, the traditional fixed costs are variable. Costs such as taxes and administrative costs are variable, since the time period varies. On the other hand, the typical traditional variable costs of labor and materials become semivariable because the materials portion is fixed when the production level is fixed while the labor would still vary with time. The revenue/cost data is different when production time, instead of production quantity, is variable; the corresponding data for the example problem is in Table 1.3.

Table 1.3 Cost/Revenue Data for Break-Even Points for Fixed
Production Quantity

Item	$/hr	Dollars ($)	Decimal
Revenue		10,000	
Fixed Costs	25.00		
Variable Costs	12.00	2,000	
Semivariable Costs	13.00	400	
Required Return		600	
Tax Rate			0.40

The *shutdown point* is found by:

revenue = semivariable costs + variable costs
$$10,000 = 13.00y + 400 \quad + 12.00y + 2000$$
$$10,000 = 25y + 2400$$
$$7,600 = 25y$$
$$y = 304 \text{ hr} \tag{1.5}$$

The *break-even at cost* is found by:

revenue = total costs
revenue = semivariable costs + variable costs + fixed costs
$$10,000 = 13.00y + 400 \quad + 12.00y + 2000 + 25.00y$$
$$10,000 = 50y + 2400$$
$$7,600 = 50y$$
$$y = 152 \text{ hr} \tag{1.6}$$

The *break-even at required return* is found by:

revenue = total costs + required return
$$10,000 = 50y + 2400 + 600$$
$$10,000 = 50y + 3000$$
$$7,000 = 50y$$
$$y = 140 \text{ hr} \tag{1.7}$$

The *break-even at required return after taxes* is found by:

revenue = total costs   + required return/(1 − tax rate)

$$10,000 = 50y + 2400 + 600/(1.00 - 0.40)$$
$$6,600 = 50y$$
$$y = 132 \text{ hr} \tag{1.8}$$

The four break-even points are illustrated in Fig. 1.6, which is a plot of the costs and revenues as a function of the production time period. Since the production time period of 100 hours is less than any of the break-even points, the project is profitable, as indicated in Table 1.4. The lower the time period, the more profitable the project. Although the net profits of Table 1.4 are identical to those of Table 1.2, the graphs in Figs. 1.4 and 1.6 appear quite different.

The approach of the fixed time period and variable production quantity has been the standard approach of break-even analysis; but because of the associated high fixed costs of administration, the variable time approach is

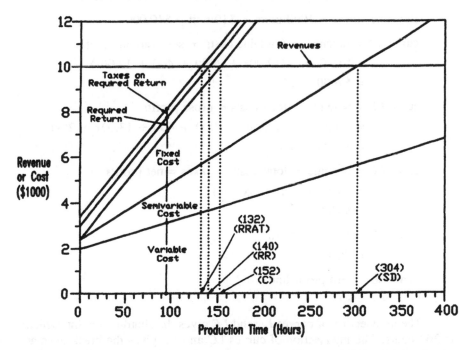

Figure 1.6 Break-even points as a function of production time period for a fixed production level.

Table 1.4 Revenues, Costs, Taxes and Profits for a Production Time Period of 100 Hours

Revenues		$10,000
Costs		
Fixed (25 × 100)	2500	
Variable (12 × 100 + 2000)	3200	
Semivariable (13 × 100 + 400)	1700	
Total Costs		7,400
Gross Profits		2,600
Taxes (@ 40%)		1,040
Net Profits		1,560

gaining more acceptance. The costs of production delays are more apparent as well as the need for more control upon administrative costs.

Another application of the variable time approach is with a profitability curve diagram. This is illustrated in Fig. 1.7 utilizing the data of Table 1.3 and a production quantity of 1000 units. The curves of Fig. 1.7 are:

Revenue Curve = 1000 units × $10/unit = $10,000

curve LA = revenue − (variable costs + semivariable costs)

$\qquad$ = $10,000 − (12.00$x$ + 2000 + 400 + 13.00x)

$\qquad$ = $7600 − 25$x$

curve LB = revenue − total costs = gross profits

$\qquad$ = $10,000 − (12.00$x$ + 2000 + 400 + 13.00x + 25x)

$\qquad$ = $7600 − 50$x$

curve LC = revenue − total costs − taxes = net profits

$\qquad$ = (curve LB)(1 − tax rate)

$\qquad$ = $4560 − 30$x$

curve LD = required return

$\qquad$ = $600

curve LE = zero-profit line

$\qquad$ = $0

The intersection of curves LA and LE gives the shutdown point, which is 304 hours. The intersection of curves LC and LE gives the break-even at cost, which is 152 hours. The intersection of curves LB and LD gives the break-even at required return, which is 140 hours. Finally, the intersection

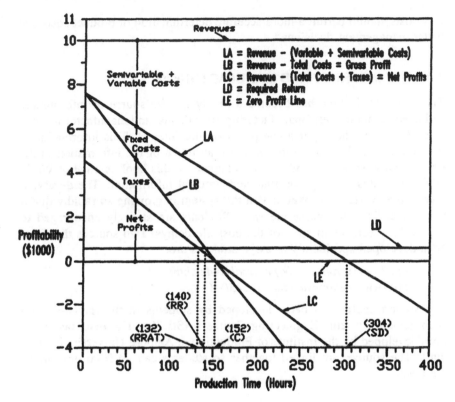

Figure 1.7   Profitability and break-even points as a function of production time.

of curves *LC* and *LD* gives the break-even at required return after taxes, which is 132 hours. Other items of importance, such as the net profits (or losses), is the difference between curves *LC* and *LE*. The gross profits (or losses) is the difference between curves *LB* and *LE* for a specific production time. The profitability curves show the importance of reducing production time for increasing profitability. The production manager has some control over the production time but little, if any, control over the production quantity. Thus, production personnel need the break-even analysis based upon production time rather than upon production quantity.

An important aspect of Figs. 1.6 and 1.7 is that fixed costs, as traditionally considered in the variable production case, are not fixed when production time is a variable. This indicates that a new approach for the evaluation of fixed, variable, and semivariable costs will be required to determine the break-even points properly when production time is a variable. These new costs should permit a more accurate determination of product costs and better management decisions.

1.4 THE INTERNET AND MANUFACTURING

The World Wide Web has become a nearly infinite source of information, and most manufacturers, manufacturing consultants, and manufacturing software developers have put home pages on the Internet. This information is updated frequently and is more recent than much of the information in the typical textbook. At the end of most chapters in this book is a list of World Wide Web sites, to help you find supplemental information. The disadvantage of the World Wide Web is that the system is growing so rapidly that it is often difficult to access the sites. Students are strongly encouraged to utilize these sources in spite of the possible delays. Two sources that have been developed specifically to assist students are:

> *http://www.iprod.auc.dk/procesdb/index.htm*
> *http://www.cemr.wvu.edu/~imse304/*

The second source has been developed by students in the Industrial and Management Systems Engineering Course IMSE 304. The first source has been developed by the Institute of Production at Aalborg University in Denmark. The sources are linked together so students can go from one site to the other.

1.5 PROFESSIONAL SOCIETIES

There are numerous professional societies concerned with various areas of manufacturing. These professional societies focus on a wide variety of aspects, including general manufacturing, engineering disciplines, special products, special processes, materials, and special services. Some of these societies are: Society of Manufacturing Engineers, Institute of Industrial Engineers, American Society for Mechanical Engineers, Society of Automotive Engineers, American Foundrymen's Society, ASM International (formerly the American Society for Metals), and AACE International (Association for the Advancement of Cost Engineering). Many other professional societies can be found on the Internet, and they often provide information and services to assist in solving problems.

1.6 SUMMARY

The importance of manufacturing to the economic success of a nation cannot be overestimated. The changes throughout the world in the last 20 years have been the result primarily of economic power, and manufacturing is the key strength of most major economies. Manufacturing has provided the jobs and incomes for the major economic powers.

Many new philosophies as well as computerized tools and techniques have been developed to assist in manufacturing. Manufacturing must be competitive not only in cost but in quality, delivery, and product aesthetics and performance. Total quality management, concurrent engineering, computer-aided process planning, group technology, and feature-based design are some of the new methodologies, philosophies, and tools to improve product productivity, quality, performance, and cost reduction. The Internet has become a major source of information for manufacturers to describe their capabilities and products.

Cost is one of the major criteria used by consumers in selecting products, and break-even analysis is one of the techniques used by producers to determine the competitiveness of a product. Two approaches to break-even analysis, the variable quantity approach and the variable time period approach, were presented. Four different break-even points were presented for evaluation: the shutdown point, the break-even point at cost, the break-even point at required return, and the break-even point at required return after taxes. Which point is most critical depends upon various factors, such as the state of the economy, the financial state of the particular company, and the production requirements of the specific department at the time of evaluation.

Manufacturing is an integrated discipline involving design, materials, processes, assembly, and management. It offers an exciting and rewarding career, but it is highly competitive. Sporting events are competitive for a season, but manufacturing is competitive year in and year out, and the pressure is constant. However, careers in manufacturing can last a lifetime, and there are always new and challenging problems.

1.7 EVALUATIVE QUESTIONS

1. What is manufacturing?

2. Define the four different break-even points.

3. Rework the example problem and determine the break-even production levels for the four break-even points for fixed costs that are $2800 instead of $2000. Also, do the revenue, costs, taxes, and profits summary as indicated in Table 1.2.

4. Rework the example problem with the fixed production quantity and determine the break-even production times for the four break-even points for

fixed costs that are $35.00 per hour instead of $25.00 per hour. Also, do the revenue, costs, taxes, and profits summary as indicated in Table 1.4.

5. Plot the profitability diagram for Evaluative Question 4.

6. The following cost/revenue data was obtained for evaluation of the various shutdown and break-even points.

Item	$/hr	Dollars ($)	Decimal
Revenue		20,000	
Fixed Costs	20.00		
Variable	10.00	2,000	
Semivariable	5.00	3,000	
Required Return		4,000	
Tax Rate (35%)			0.35

a. What is the shutdown point, in hours? (1000)
b. What is the break-even point at cost, in hours? (428.6)
c. What is the break-even point at required return? (314.3)
d. What is the break-even point at required return after taxes? (252.8)
e. Plot the profitability diagram for this problem.

7. The following cost/revenue data was obtained for evaluation of the various shutdown and break-even points. The expected production quantity is 1000 units.

Item	$/hr	Dollars ($)	Decimal
Revenue		35,000	
Fixed Costs	20.00		
Variable	7.00	2,000	
Semivariable	3.00	3,000	
Required Return		4,000	
Tax Rate (35%)			0.35

a. What is the shutdown point, in hours? (3000)
b. What is the break-even point at cost, in hours? (1000)
c. What is the cost per unit at the break-even cost point? ($35.00)
d. What is the break-even point at required return after taxes? (794.7)
e. Plot the profitability diagram for this problem.

8. Define the following terms:

    a. Concurrent engineering
    b. Group technology
    c. Total quality management
    d. Statistical process control
    e. ABC costing

1.8 RESEARCH QUESTIONS

1. Do an in-depth report on one of the new aids in manufacturing.

2. Evaluate the ABC costing approach of accounting, and compare it with manufacturing-cost estimating methods.

3. Why is the primary reason for the failure of engineering projects the result of poor management?

4. Visit one of the Web sites listed at the end of this chapter.

REFERENCES

1. Schey, John A. Introduction to Manufacturing Processes, 2nd ed., 1987, Mc-Graw-Hill, New York, p. 6.
2. World Tables. John Hopkins University Press, Baltimore, 1995.
3. Winner, R. I., et al. 1988, The Role of Concurrent Engineering in Weapons Systems Acquisition, IDA Report R-338, December, Institute for Defense Analysis, Alexandria, VA.
4. Creese, R. C., and Ham, I. 1979. "Group Technology for Higher Productivity and Cost Reduction in the Foundry," AFS Transactions, vol. 87, pp. 227–230.
5. Postula, Frank D. 1989. "Total Quality Management and the Estimating Process-A Vision." Paper for BAUD 653, Systems Acquisition and Project Management, July 13.
6. Chang, T. C., and Wysk, R. A. An introduction to Automated Process Planning Systems, 1985, Prentice-Hall, Englewood Cliffs, NJ, p. 214.
7. Ravi, B. Manufacturability Analysis of Cast Components, Ph.D. Thesis, Indian Institute of Science, Bangalore, India, 1992.
8. Creese, Robert C. "Break-even Analysis—The Fixed-Quantity Approach," 1993 AACE Transactions, AACE, Morgantown WV, pp. A1.1–A1.7.
9. "Group Technology." Computer Technology vol. 9, no. 2, Oct. 1987, pp. 83–91.

10. Torino, John. "Making It Work Calls for Input from Everyone," IEEE Spectrum, vol. 28, July 1991, p. 31.

INTERNET SOURCES

Aalborg University Process Database: *http://www.iprod.auc.dk/procesdb/index.htm*

Industrial & Management Systems Engineering: *http://www.cemr.wvu.edu/~imse304/*

Professional Organizations: *http://www.ems.psu.edu/Metals/features/profsoc.html*

II
MATERIAL AND DESIGN CONSIDERATIONS IN MANUFACTURING

2

Basic Material Properties

2.1 INTRODUCTION

The integration of design, materials, and processing has always been recognized as an important consideration, but the focus has been primarily upon
a sequential rather than concurrent approach. The material was selected, the
design was made with the selected material, and the material was processed
to give the desired shape. In a concurrent environment, the design function
is related to certain desired material properties, the materials are selected
according to the desired properties, and then the final design is developed
with respect to the material properties and the material processing required
to obtain the desired design material properties. Two of the basic items that
provide a start at examining material properties with respect to design and
processing are the material crystalline structure and the material phase
diagram.

A basic understanding of material properties is essential to understand
how materials are selected by designers and how processing can affect the
properties of materials. The emphasis will be upon the structures of metals,
because more products are manufactured with metals and more information
is available about metals than about other materials.

2.2 ATOMIC BONDING

The three primary types of atomic bonding are ionic, covalent, and metallic.
In *ionic bonding*, the valence electrons are given up by atoms that have only
a few valence electrons, usually one or two, and taken by the atoms that
have a nearly complete outer shell and need only one or two electrons to
complete the outer shell. A typical ionically bonded material is NaCl, or
salt, where the sodium atom (Na) gives up its valence electron to complete
the outer shell of the chlorine (Cl) atom. Ionic materials are generally very
brittle, and very strong forces exist between the two ions.

25

With *covalent bonding*, the valence electrons are shared between two particular atoms. An atom may share electrons with more than one atom, and many compounds have covalent bonding, such as H_2, CO_2, and SiO_2. Polymer structures typically are long chains of covalently bonded carbon and hydrogen atoms in various arrangements.

Metallic bonding occurs when the valence electrons are not associated with a particular atom or ion, but exist as an "electron cloud" around the ion centers. The metallic bonding permits easy movement of the electrons, so materials with metallic bonding have good electrical and thermal conductivity when compared to materials with covalent or ionic bonding. Metals are materials that predominantly have metallic bonding.

Materials generally do not have pure metallic, pure ionic, or pure covalent bonding; they are predominantly one type of bonding but may have some other forms of bonding as well. For example, iron is predominantly a metallically bonded material, but some covalent bonding also occurs. Thus when a material is said to have metallic bonding, it implies that the *dominant* type of bonding is metallic bonding and not that metallic bonding is the only type of bonding that occurs in that material.

The van der Waals bond occurs in all materials, but it is especially important in covalently bonded materials, for this bonding is what holds the molecules together. This bonding is critical for plastics and polymers and is considered a primary bond for those materials. When ionic or covalent bonding occurs, there is still some imbalance in the electrical charge of the molecule. The imbalance in the charge creates forces of attraction that bond the molecules together. This bonding is the van der Waals bonding.

2.3   CRYSTALLINE STRUCTURE

Materials can generally be classified (2) as either amorphous or crystalline. Amorphous materials have no long-range order; that is, the atoms or molecules are not periodically located over long distances. Amorphous materials usually have short-range order that indicates where the nearest neighbor atom is; many solids, such as glasses and polymers, are amorphous materials. Crystalline materials have both long-range and short-range order; that is, if the precise atom arrangement in one position is known, then the atom arrangement in another position can be predicted exactly. Some metals— that is, those that have been cooled extremely quickly—form an amorphous structure and will have hard and brittle properties like glasses. These have been called metal-glasses.

The crystalline structure of a material consists of a three-dimensional arrangement of points in space where each lattice point has identical surroundings. The coordinate system associated with a crystalline lattice has three axes and three coordinate angles between the axes; this system is called a *crystal system*. There have been only seven different crystal lattice systems observed, and these are described in Table 2.1. The lattice parameters a, b, and c refer to the unit lengths along the three axes, x, y, and z, respectively. The angles (α, β, Γ) refer to the coordinate angles between the x, y, and z axes.

In addition to the seven different crystal systems, there are four different unit cell structures that have been observed. These four unit cell structures are the simple, the base-centered, the face-centered, and the body-centered structures. A corner atom can be considered to be one-eighth in a unit cell, since it would be in eight different unit cells. Similarly, a face atom can be considered to one-half in a unit cell, and a center atom is entirely in the unit cell. A rectangular cell would have eight corners, six faces, and one center. In the base-centered structure, only two opposite faces, the top and the bottom of the six faces, have atoms in the center of the faces as well as at the corners; the four side faces do not have atoms in the center of the faces. Sketches of two different cells are illustrated in Figure 2.1. The number of atoms per unit cell for the different unit cells is shown in Table 2.2.

If one considers the different combinations of crystal lattice systems and unit cells, one would expect 28 different combinations. However, there is some redundancy and only 14 different combinations have been observed; these are called the *Bravais lattices*. The 14 different Bravais lattices are presented in Table 2.3 in terms of crystal structure and unit cell description for each lattice.

The most common Bravais lattices that are observed among metals are: the cubic—body-centered, commonly called the *body-centered cubic*

Table 2.1 The Seven Different Crystal Lattice Systems

System name	Lattice parameters	Angle relationships
Cubic	$a = b = c$	$\alpha = \beta = \Gamma = 90°$
Tetragonal	$a = b \neq c$	$\alpha = \beta = \Gamma = 90°$
Orthorhombic	$a \neq b \neq c$	$\alpha = \beta = \Gamma = 90°$
Rhombohedral	$a = b = c$	$\alpha = \beta = \Gamma \neq 90°$
Hexagonal	$a = b \neq c$	$\alpha = \beta = 90°, \Gamma = 120°$
Monoclinic	$a \neq b \neq c$	$\alpha = \beta = 90°, \Gamma \neq 90°$
Triclinic	$a \neq b \neq c$	$\alpha \neq \beta \neq \Gamma \neq 90°$

a, b, and c refer to the unit lengths along the x, y, and z axes, respectively. α, β, and Γ refer to the coordinate angles between the axes.

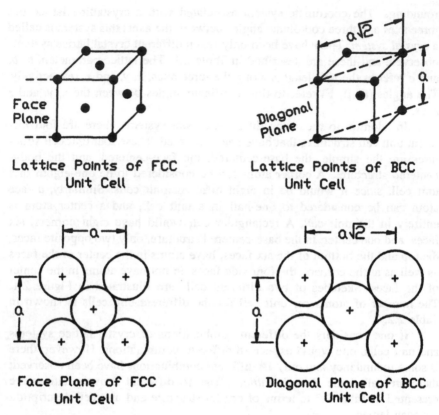

Figure 2.1 Unit cell structure for face-centered cubic (FCC) and body-centered cubic (BCC) unit cells.

Table 2.2 The Number of Atoms per Unit Cell for the Four Different Unit Cells

Unit cell name	Number of Atoms per unit cell
Simple	1
Base-centered	2
Face-centered	4
Body-centered	2

Table 2.3   The Fourteen Different Bravais Lattices

Crystal System and Unit Cell	Crystal System and Unit Cell
1. Cubic—face-centered	10. Monoclinic—simple
2. Cubic—body-centered	11. Monoclinic—base-centered
3. Cubic—simple	12. Triclinic—simple
4. Tetragonal—body-centered	13. Hexagonal—simple
5. Tetragonal—simple	14. Rhombohedral—simple
6. Orthorhombic—body-centered	
7. Orthorhombic—base-centered	
8. Orthorhombic—face-centered	
9. Orthorhombic—simple	

(*BCC*); the cubic—face-centered, commonly called the *face-centered cubic* (*FCC*); and the simple hexagonal, commonly called the *hexagonal close-packed* (*HCP*). Some of the body-centered materials at room temperature are chromium, manganese, iron, tungsten, and vanadium. Materials that are face-centered cubic are silver, gold, aluminum, copper, nickel, and lead. Some materials, such as iron, change crystal structure depending upon the temperature; these are called *allotropic* materials. Iron in the face-centered cubic structure is called *austenite*, whereas iron at room temperature in the body-centered cubic structure is called *ferrite*. The structure has an important effect upon the material properties; thus, the austenite irons have different material properties than the ferrite irons.

The material property of density is related to its structure. The density for a cubic unit cell can be determined by:

$$\rho = M/V \tag{2.1}$$

$$= \frac{MW/\text{Å} \times N}{a^3}$$

$$= \frac{MW \times N}{\text{Å} \times a^3} \tag{2.2}$$

where

ρ = density (g/cm^3)
M = mass (g)
V = volume (cm^3)
MW = molecular weight (g/g mole)
N = atoms per unit cell
a = lattice parameter (cm)
Å = Avagadro's mumber (6.02 × 10^{23} atoms/g mole)

If the lattice parameter for the body-centered cubic structure of iron is 2.86 angstroms (1 angstrom = 1×10^{-8} cm) and the molecular weight is 55.85, what is the density of iron? Using Eq. (2.2) and the fact that there are two atoms per unit cell of a BCC structure, the value would be:

$$\rho = \frac{55.85 \text{ g/g-mole} \times 2 \text{ atom/unit cell}}{6.02 \times 10^{23} \text{ atoms/g-mole} \times (2.86 \times 10^{-8} \text{ cm})^3}$$
$$= 7.93 \text{ g/cm}^3$$

The density of the unit cell is the theoretical density of the material; the actual density varies, since the material has vacancies and impurities. The theoretical density is usually within 2 percent of the actual density of the pure material.

The dimensions of the atoms and the packing density can also be determined from the lattice parameter and the unit cell structure. The face-centered cubic and body-centered unit cell structures are shown in Fig. 2.1. From the lattice parameter of iron and the body-centered cubic structure, the length of the diagonal of the cube is two atom diameters, so:

$$\text{Diagonal} = 2d = a\sqrt{3}$$

or

$$d = a\sqrt{3}/2 \qquad\qquad\qquad (2.3)$$

where

d = atom diameter
a = lattice parameter

Thus:

$$d = 2.86\sqrt{3}/2$$
$$= 2.48 \text{ angstroms}$$

The density of packing is that portion of the unit cell that is filled with atoms. If the atoms are considered to be solid spheres, the density of packing for a body-centered unit cell is:

$$\rho_p = \frac{\text{number of spheres} \times \text{volume of spheres}}{\text{cell volume}} \qquad\qquad (2.4)$$

For the body-centered cubic unit cell, the density of packing is:

$$\rho_p = \frac{2\pi d^3/6}{a^3} \qquad\qquad\qquad (2.5)$$

where

ρ_p = density of packing
d = atom diameter
a = lattice parameter

For the BCC unit cell, $a = 2d/\sqrt{3}$, so:

$$\rho_p = \frac{2 \times \pi d^3/6}{(2d/\sqrt{3})^3}$$
$$= \pi\sqrt{3}/8$$
$$= 0.68$$

The face centered cubic unit cell (which is a close-packed structure) and the hexagonal close-packed structure have density of packing values that are equal and are 0.74. The close-packed structures also have close-packed planes, and slip can occur more easily on the close-packed planes. Thus, in general, materials that have close-packed structures deform more easily than do body-centered cubic materials that are not close-packed. Thus, since deformation is more difficult to occur, the body-centered cubic materials tend to have higher strength values than do the close-packed materials. The face-centered cubic materials have the most slip systems and thus are the easiest to deform without fracture. Thus, in deformation processing, materials with face-centered cubic structures tend to have better formability than materials with body-centered or hexagonal closed-packed structures.

Although the density of packing for the face-centered cubic structure is greater than the density of packing for the body-centered cubic structure, the void spaces in the FCC structure can be larger than those of the BCC structure. This is evident in iron, where austenite, which is FCC, can dissolve up to 2 percent carbon, whereas ferrite, which is BCC, can dissolve only a maximum of 0.025 percent carbon. Thus, although the density of packing is greater in the FCC structure, the hole sizes can be greater than those in the BCC structure. There are fewer holes in the FCC structure, and the total void volume is less than that of the BCC structure.

2.4 MILLER INDICES

The Miller indices are used to describe planes and directions within the crystal structure. A *slip system* is defined as a slip direction within a slip plane. The close-packed planes of the FCC and HCP systems are further apart then the nonclose planes; thus, slip can occur more easily. In the FCC structure, there are 12 close-packed slip systems. A *close-packed slip system* is defined as a close-packed plane and a close-packed direction. In the HCP system, there exists only three close-packed slip systems, whereas there are no close-packed slip systems in the BCC structure. The more easily slip occurs, the easier it is to deform the material. Conversely, materials that do not slip easily tend to have high tensile strengths.

The Miller indices for a direction are represented by the nomenclature $[u,v,w]$, where the brackets indicate it is a direction and the values of u, v, and w indicate the magnitude and direction of the vector from the origin in the respective x, y, and z directions. For example, the direction [3,2,2] is indicated in Fig. 2.2a. The values 3,2,2 represent the number of unit lengths in the x, y,and z directions. In the cubic system the unit lengths are the same for all three directions, but the Miller indices work for all crystal systems, not only the cubic system. If a direction had the Miller indices of [6,4,4], this would indicate that the vector is in the same direction as the [3,2,2] vector. Since the purpose of the Miller indices is to indicate only the direction and not the magnitude, these two directions are considered equivalent. Parallel vectors will also have the same Miller indices for direction.

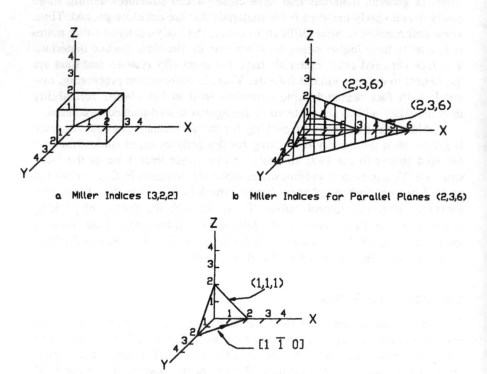

a Miller Indices [3,2,2]   b Miller Indices for Parallel Planes (2,3,6)

c Slip System with (1,1,1) Plane and [1 $\bar{1}$ 0] direction

Figure 2.2 Miller indices and slip systems.

The Miller indices for a plane are represented by the nomenclature of (h,k,l), where the parentheses indicate that it is a plane and the values of h, k, and l relate the intercepts of the plane to the axis system. Planes that have an integer multiple of the Miller indices of a plane are parallel planes and can be considered as the same plane. The two planes in Fig. 2.2b are parallel and have the same Miller indices. The Miller indices for a plane are the least-integer set of numbers related to the reciprocal of the intercepts of the plane. For example, in Fig. 2.2b, the intercepts for the inner plane are 3,2,1. The Miller indices of the plane are derived by finding the reciprocal of the intercepts in the lowest-integer set; that is:

Intercepts	3	2	1
Reciprocal of intercepts	1/3	1/2	1
Multiply by 6 to obtain all integers	2	3	6
Miller indices	(2,3,6)		

Note that if we consider the intercepts of the outer plane, we will obtain the same values for the Miller indices. To sketch a particular plane, we must reverse the procedure to determine the intercepts.

In Fig. 2.2c, the slip system shown contains the $[1,-1,0]$ direction and the $(1,1,1)$ plane. Note that the Miller Indices can be negative for both the plane and the direction. To find the direction in the plane, it is easiest to draw the parallel vector from the origin, determine its direction, and then determine the parallel direction in the plane, as shown. There are special relationships between the Miller indices for planes and directions in cubic systems, and thus most applications and examples are for cubic systems. Some modifications are necessary for the hexagonal system, and four values are used to describe the planes and directions instead of three. Further details can be found in most basic books on materials engineering or materials science.

2.5 PHASE DIAGRAMS

Phase diagrams are temperature-composition plots for alloys, and many of the diagrams used are binary diagrams. A *binary diagram* is for two elements, such as iron and carbon, which make up the iron carbon phase diagram. The purpose of the diagram is to tell what phases are present for specific binary compositions at specific temperatures. Although the phase diagram is applicable only for equilibrium conditions, it does provide a basis for predicting what would occur under nonequilibrium conditions. An understanding of phase diagrams is necessary for evaluating materials for processing, for they can be used to predict characteristics such as superheat, fluidity, crystal structure, coring, castability and formability.

The terminology of phase diagrams is essential to understanding what is occurring as temperature changes. Some of the important terms follow.

Phase: A homogeneous, physically distinct substance; a constituent that is completely homogeneous both physically and chemically, i.e., the same chemical composition, lattice parameter, and unit cell structure.

Component: A distinct part of the ingredients. A binary system has two components, and they are the two elements used in the system.

Phase diagram: A graphical representation of temperature and composition limits of the phase fields in an alloy as they exist under the specific equilibrium conditions of heating and cooling.

Liquidus line: The line that indicates the temperature at which the first solid appears upon cooling or the temperature at which the last solid disappears upon heating. In a few instances where a liquid separates into two liquids, the temperature at which the separation starts is also called a liquidus line.

Solidus line: The line that indicates the temperature at which the last liquid disappears upon cooling.

Solvus line: The line that indicates the solubility limits.

Invariant reaction line: The line connecting the three phases that are in equilibrium at a specific temperature in the binary system. It appears as a horizontal line because it is at constant temperature.

Figure 2.3a represents a generic general phase diagram. The numbers represent points where the curves intersect, so they can be used to represent the various lines. The two components are A and B, and the percentages are expressed as weight percent B. Since it is a binary diagram, 80 percent B implies that the alloy is 80 percent B and 20 percent A; that is, the sum of the two components must be 100 percent. With respect to the definitions of the various lines, line 10-11 is a solidus line, line 9-11 is a liquidus line, and line 10-14 is a solvus line. Line 10-11-12 is an invariant reaction line at a temperature of 1800°C, at which the phases sigma (σ), liquid (L), and delta (δ) are in equilibrium. The invariant reaction is written for cooling; that is, the invariant reaction is $L = \sigma + \delta$ as liquid transforms to sigma and delta upon cooling. If heating occurs, the reverse reaction would occur; that is, $\sigma + \delta = L$.

On the phase diagram, frequently only the single-phase regions are identified. In the two-phase regions, the two phases present are those that

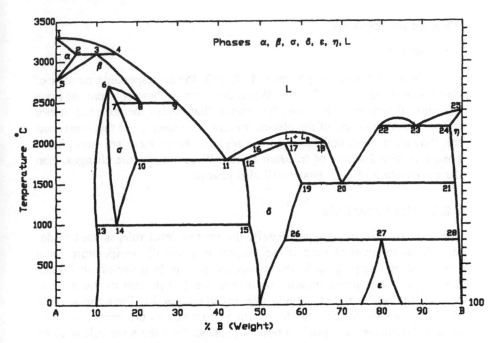

Figure 2.3a General phase diagram.

bound the region on the two sides. For example, in the region 7-8-9-11-10-7, the phases are determined by the two sides 9-11 (which is the L, or liquid, phase boundary) and 10-7 (which is the sigma (σ) boundary). Thus the two phases in the region are liquid and sigma. The composition of the phases will depend upon the temperature.

2.5.1 The Phase Rule

The phase rule is used to determine the number of degrees of freedom that exist in the system at a specific temperature and at a specific composition. For a binary system the rule is:

$$DF = C - P + 1$$

where

DF = degrees of freedom
C = number of components = 2 for a binary system
P = number of phases (excluding the gas phase)

Thus the rule becomes

$$DF = 3 - P$$

Since the number of phases is 1, 2, or 3, the corresponding number of degrees of freedom is 2, 1, or 0. When there are 3 phases present, or there are zero (0) degrees of freedom, that means that neither the temperature nor the composition of any of the phases present can change, and this represents the "invariant" reaction lines. In the regions where there are two phases present (where 1 degree of freedom exists), if the temperature changes, then the composition of each phase will also change.

2.5.2 The Lever Rule

The lever rule, or more appropriately, the reverse lever rule, is used to determine the amount of each phase present at a specific temperature for a specific alloy composition in the two-phase region. In a one-phase region, there is only one phase present, so it must be 100 percent of the amount present. In the three-phase region (the invariant reaction line), the amount of each phase is changing as the reaction proceeds, so the lever cannot be used to determine the specific amounts present. Thus the lever rule is to be applied only in the two-phase region.

If one considers an alloy of 30 percent B (and 70 percent A) at a temperature of 2000°C, as illustrated in Figure 2.3b, the phases present are sigma (σ) and liquid (L). The line xyz represents the lever to be used for the temperature of 2000°C. The composition of the sigma phase is determined by the point x and the reading on the weight percent scale that give approximately 18 percent B (82 percent A). Similarly, the composition of the liquid phase is found by reading point z on the composition scale; the value is approximately 38 percent B (62 percent A). From the compositions, the amounts of the phases can be determined by means of the lever rule. The lever is the line xyz, which consists of two parts, xy and yz. The amount of the sigma phase, represented by the composition x, is the portion of the lever yz/xyz; that is:

$$\text{Amount } x, \text{ or sigma} = \frac{yz \text{ portion of lever}}{\text{total lever } xyz \times 100}$$

For the 30 percent B alloy at 2000°C, the values are:

$$\text{Amount } x, \text{ or sigma} = \frac{38 - 30}{38 - 18} \times 100 = 40\%$$

Thus, the amount of the sigma phase present at 2000°C is 40 percent

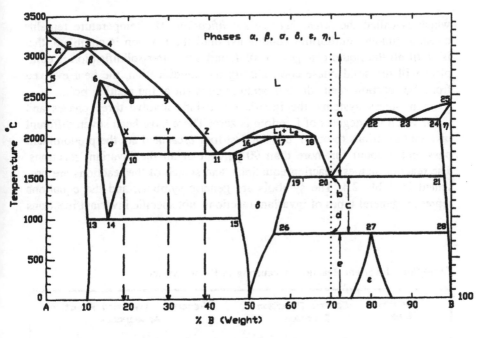

Figure 2.3b General phase diagram illustrating the lever rule.

of the total amount present. The amount of the liquid phase should be the remainder, or 60 percent. It can be calculated by:

$$\text{Amount } z, \text{ or liquid} = \frac{xy \text{ portion of lever}}{\text{total lever } xyz} \times 100$$

$$= \frac{30 - 18}{38 - 18} \times 100 = 60\%$$

Note that it is the opposite portion of the lever that determines the amount of the phase present, and that is why the rule is often called the reverse lever rule.

2.5.3 Invariant Reactions

Invariant reactions are important, because they indicate conditions where the phases are changing; some (one or two) phases will react and form one or two new phases. During the formation of the new phases, energy is evolved as heat during cooling. In a one-component system, the invariant point would be the melting point, where the liquid solidifies to a solid. During solidification, the liquid phase changes to the solid phase and heat is evolved, which is called the *latent heat of solidification*. The temperature remains constant (under equilibrium conditions) until the reaction is complete, that is, until all the liquid changes to solid, and only after solidification is complete will the solid phase cool. During the solidification, the heat evolved from the reaction keeps the temperature constant at the melting point.

In binary systems, the invariant reactions involve three phases and occur when the degrees of freedom is zero. There have been seven different invariant reactions to occur, but the first four presented are the predominant ones and account for more than 90 percent of all the invariant reactions. The reaction names, reaction equations, and sketch of the reactions are presented in Table 2.4. The symbols are generic symbols, and the equations represent general types of invariant reactions, not specific invariant reactions

Table 2.4 Invariant Reactions Occurring in Binary Systems

Reaction Name	Generic Reaction Equation	Sketch of Invariant Reaction Appearance
1. Eutectic	$L = \alpha + \beta$	
2. Eutectoid	$\sigma = \alpha + \beta$	
3. Peritectic	$\alpha + L = \beta$	
4. Peritectoid	$\alpha + \beta = \sigma$	
5. Monotectic	$L_1 = \alpha + L_2$	
6. Syntectic	$L_1 + L_2 = \alpha$	
7. Catatectic	$\alpha = \beta + L$	

on specific diagrams. The equation reactions are written for cooling conditions; the left-hand side of each equation transforms to the right-hand side of the equation upon cooling. The equations would be reversed for heating conditions.

The generic phase diagram (Fig. 2.3a) has eight different invariant reactions. The invariant reaction at 70 percent B and 1500°C, which is located at point 20 on the diagram, is a eutectic reaction. The specific eutectic reaction is:

$$L = \delta + \eta$$

Note that the region above point 20 is where the liquid phase exists and that below the line is the two-phase region where δ and η exist. Thus the liquid solidfies to the two solid phases δ and η upon cooling. The temperature remains constant until all the liquid has solidified. The composition of the liquid is 70 percent B, the composition of the δ is approximately 60 percent B, and the composition of the η phase is approximately 98 percent B.

One of the most common eutectic reactions is the formation of gray cast iron as the liquid solidifies to austenite and graphite flakes. The decomposition of the austenite to ferrite and cementite is an example of an important eutectoid reaction. Peritectic reactions occur in brass alloys, and peritectoid reactions are not very frequent. The monotectic reactions often occur in systems with elements of quite different melting points, such as lead-zinc or lead-copper systems. The syntectic and catatectic reactions occur infrequently.

2.5.4 Cooling Curves

Cooling curves are temperature–time plots of the temperature as an alloy cools. The differences in the slopes of the curve can be attributed to the heat being evolved during the solidification process. As the liquid cools, heat is being removed from the system but no heat is being generated. When solidification starts to occur, heat is evolved during solidification; and if the heat is being removed from the system at a constant rate, the cooling of the liquid will be slower. When an invariant reaction occurs, large amounts of heat may be evolved; and if the heat is generated at a rate equal to its removal rate, the alloy will remain at a constant temperature. Figure 2.4 indicates the cooling curve for an ideal eutectic system and the corresponding phase diagram. In practice, the cooling curves of various alloys are made to determine the liquidus, solidus, and invariant reaction temperatures for the phase diagram. Cooling curves can be used to predict compositions and grain sizes of alloys.

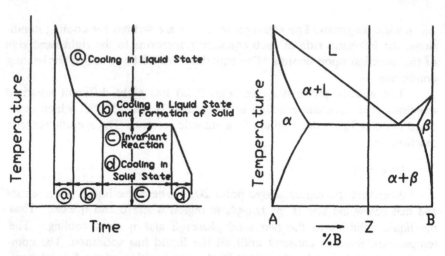

Figure 2.4 Cooling curve and corresponding phase diagram for the cooling of alloy Z.

Actual cooling curves vary somewhat from the ideal cooling curve. It is usually difficult to determine the actual liquidus temperature unless precise measuring and very slow cooling rates are used. In foundry practice, there is often a amount of undercooling that occurs, and the amount of under-cooling indicates the type of grain structure being produced (equiaxed or columnar, fine or coarse). Figure 2.5 gives an indication of the differences between the actual and ideal cooling curves. The liquidus temperature and the eutectic temperature values can be used to predict the composition fairly accurately. Many foundry operations, particularly cast iron and aluminum foundries, use special instruments to determine the liquidus and eutectic temperatures to predict the alloy composition or to determine if proper nucleation is occurring for grain size control.

2.5.5 Alloy Cooling Descriptions

The description of how an alloy cools can be useful in predicting properties of the alloy. A form for describing the cooling of an alloy has been developed and is presented in Table 2.5 for an alloy of 70 percent B that is cooled from 3000°C. Each composition of the alloy has a different cooling description, and these cooling descriptions can become quite long for alloys where numerous invariant reactions occur. Note that at the invariant temperatures, the alloy is at a fixed temperature until the reaction is complete; only then can further cooling occur. Thus Table 2.5 illustrates the five different ranges

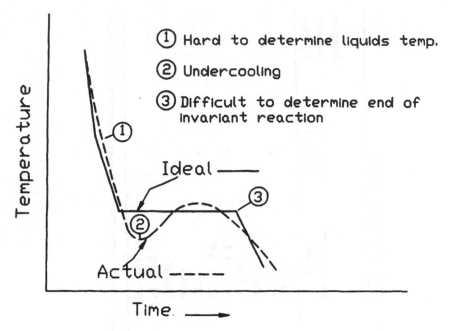

Figure 2.5 Comparison of actual and ideal cooling curves.

or invariant values of importance in the cooling from 3000°C to room temperature (RT).

The alloy cooling descriptions (ACDs) provide information about how the phases were formed, and not just what phases were formed. The ACDs indicate what phases should be present and the amounts of the phases, but not whether the structure formed is fine or coarse. The ACDs are used to predict the equilibrium structures, but they can also be used to predict nonequilibrium structures, for some reactions may not occur or not go to completion at rapid cooling rates.

Some basic information can be obtained from the alloy cooling description. For example, the range in which the phase is a single liquid indicates the amount of superheat. In the example in Table 2.5, the alloy is all liquid from 3000° to 1500 degrees°, and thus the superheat is the temperature difference: 3000 − 1500 = 1500°. Superheat can also be described as the amount of temperature above the liquidus temperature. The freezing range of the alloy is the range between the liquidus and solidus lines; for the alloy selected, the freezing range is zero, since the alloy has a single freezing point at the eutectic temperature. If the alloy was 65 percent B, the freezing range would be from 2000° to 1500°, or 500°.

Table 2.5 Alloy Cooling Description for 70% B Generic Alloy from 3000°C to Room Temperature

Alloy temperature range or point (symbol); see Figure 2.3b	Cooling temperature range or value (value °C)	Phases present	Composition of phases	Degrees of freedom	Invariant reaction equation
a	$3000 > T > 1500$	L	70% B	2	—
b	$T = 1500$	L	70% B	0	$L = \delta + \cap$
		δ	60% B		
		$\cap$	98% B		
c	$1500 > T > 800$	δ	60–55% B[a]	1	—
		$\cap$	98–99% B		
d	$T = 800$	δ	55% B	0	$\delta + \cap = \varepsilon$
		$\cap$	99% B		
		ε	80% B		
e	$800 > T > RT^{b}$	δ	55–50% B	1	—
		ε	80–75% B		

[a] The composition changes during the temperature range; the first value corresponds to the composition at the upper temperature, and the second value corresponds to the composition at the lower temperature.

[b] RT stands for "room temperature."

2.5.6 The Iron Carbon Phase Diagram

The iron carbon (Fe–C) phase diagram is the most important phase diagram, because iron and steel are the most common alloys used in manufacturing. The most important region of the diagram is where the carbon is less than 7 percent. This portion of the diagram is also referred to as the iron–iron carbide diagram, since iron carbide is the substance that is usually found at room temperature rather than iron and carbon. A sketch of the iron carbon phase is provided in Figure 2.6. Although it is an equilibrium phase diagram, it is very helpful in understanding the nonequilibrium reactions that occur in heat treatment.

There are actually two phase diagrams in Figure 2.6; the iron–iron carbide (Fe–Fe$_3$C) diagram and the iron–graphite (Fe–C) diagram. The iron–iron carbide diagram is indicated by the solid lines, and the carbon is in the form of iron carbide, represented by Fe$_3$C. Iron carbide is also called cementite. The iron–graphite diagram is very similar, but the differences are indicated by the dashed lines, that is, when graphite is the second phase

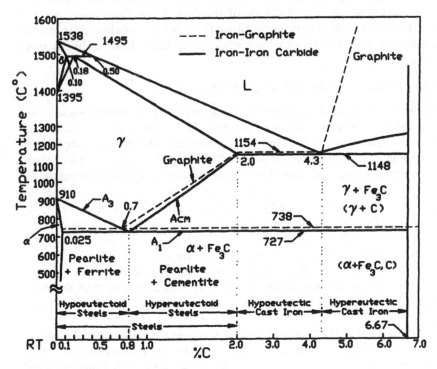

Figure 2.6 Iron carbon phase diagram.

formed instead of iron carbide. For high-carbon alloys, such as cast irons, graphite usually forms during the eutectic reaction (4.3 percent C) instead of iron carbide. On the other hand, for low-carbon alloys, such as steels, the eutectoid reaction (0.8 percent C) leads to the formation of pearlite [ferrite and cementite (Fe_3C)] instead of ferrite and graphite.

The iron–iron carbide diagram indicates that steels are regarded to have carbon contents less than 2 percent, and cast irons have carbon contents greater than 2 percent. The scale is expanded from 0 to 2 percent to indicate more clearly the temperatures and composition values. Note that there are three invariant reactions: eutectic, peritectic, and eutectoid. The eutectic reaction is extremely important for cast irons, and the eutectoid reaction is extremely important in understanding steels. The peritectic reaction is important in the formation of hot tears in steel castings. The iron–iron carbide phase diagram is also useful in understanding the transformation diagrams in heat treatment of steels.

Other elements will cause changes in the phase diagram. If another element, such as silicon, is added to the system, the invariant reaction lines are no longer required to be constant, because the additional component has provided an additional degree of freedom. High alloy steels, such as stainless steels, which have approximately 20 percent of chromium, nickel, and manganese in total, can have austenite (gamma iron) stable at room temperature.

2.5.7 Microconstituents and the Lever Rule

The lever rule is used to determine the amounts of the phases present, but it can also be used to determine the amounts of the microconstituents. A *microconstituent* is the structure one would see with a microscope when examining a properly prepared polished and etched sample. For most cases, the microconstituent is formed when the eutectic or eutectoid reaction occurs. For example, the eutectoid structure on the iron–iron carbide phase diagram in Figure 2.6 is pearlite, which is a mixture of cementite (Fe_3C) and ferrite. The material properties are determined by the amounts of pearlite and ferrite, rather than the amounts of ferrite and cementite. The compositions of the phases are 0.025 percent C for the ferrite, 6.67 percent C for the cementite, and 0.8 percent carbon for the microconstituent pearlite. Since the carbon content of the ferrite is so much smaller than the other carbon amounts, it will be treated as zero (0.00) for the following calculations.

For a steel of 0.20 percent carbon, estimate the amounts of the phases ferrite and cementite and the amounts of the microconstituents, ferrite and pearlite. From the reverse lever rule, the amounts of the phases would be:

$$\% \text{ ferrite} = (6.67 - 0.20)/(6.67 - 0.00) \times 100 = 97.0\%$$
$$\% \text{ cementite} = (0.20 - 0.00)/(6.67 - 0.00) \times 100 = 3.0\%$$

Thus the total ferrite is 97% and the total cementite is only 3.0% of the mixture.

The amounts of the microconstituents would be:

$$\% \text{ ferrite (free)} = (0.80 - 0.20)/(0.80 - 0.00) \times 100 = 75\%$$
$$\% \text{ pearlite} = (0.20 - 0.00)/(0.80 - 0.00) \times 100 = 25\%$$

Thus, if we looked at the structure under a microscope, we would see a mixture of 75% ferrite and 25% pearlite. The question is what happened to the other 22% of the ferrite? Pearlite is a mixture of the two phases, ferrite and cementite. It can be determined to be:

$$\% \text{ ferrite (in pearlite)} = (6.67 - 0.80)/(6.67 - 0.20) \times 100 = 88.0\%$$
$$\% \text{ cementite (in pearlite)} = (0.80 - 0.20)/(6.67 - 0.20) \times 100 = 12.0\%$$

Thus the total amount of ferrite would be the free ferrite plus the ferrite in the pearlite; that would be:

$$\% \text{ ferrite (total)} = \text{free ferrite} + \text{pearlite ferrite}$$
$$= \% \text{ ferrite (free)} \times \text{amount of ferrite in free ferrite}$$
$$+ \% \text{ pearlite} \times \text{amount of ferrite in pearlite}$$
$$= 75\% \times 1.00 + 25\% \times 0.88$$
$$= 97\%$$

Thus to predict mechanical material properties, the key numbers would be the amounts of ferrite (75%) and pearlite (25%), which are the microconstituents, and not the total amount of the phases of ferrite (97%) and cementite (3%). As discussed in Chapter 3, the volume amounts are the values to be used rather than the weight amounts, but for steels they are approximately the same.

In an aluminum silicon alloy, the amount of the eutectic structure and the associated primary phase are important, rather than the amounts of the two phases. In determining the amounts of the microconstituents, the lever rule calculations are made at the temperature at which the invariant reaction occurs rather than at room temperature or any other temperature at which the microconstituents exist.

2.6 SUMMARY

The crystal structure and equilibrium phase diagrams are two basic types of material information that are very useful in materials selection and processing. Materials with the same crystal structure often have similar properties; for example materials with the FCC structure are often easy to form because they have more close-packed slip systems, whereas materials with the BCC structure tend to be high-strength materials. The phase diagrams are useful to predict the casting properties, such as freezing range and superheat, and to predict the material properties from the amounts and types of phases present. These basic types of information are often very useful in finding alternative or substitute materials.

2.7 EVALUATIVE QUESTIONS

1. Sketch a face-centered unit cell, and show the calculations for the number of atoms in the face-centered unit cell.

2. Calculate the theoretical density of nickel if the lattice parameter is 3.52 angstroms and the molecular weight is 58.71.

3. Calculate the diameter of the nickel atom if the lattice parameter is 3.52 angstroms and the unit cell is face-centered.

4. Show that the density of packing for the face-centered cubic unit cell is 0.74. Also calculate the density of packing for the simple cubic unit cell.

5. Iron changes from the face-centered cubic structure, which is called austenite, to the body-centered cubic structure, which is called ferrite, when cooling in the solid state. If the atom diameter of iron is 2.5 angstroms, calculate the volume of one gram mole in both the FCC structure and the BCC structure. Does it expand or contract?

6. What are the names of the four different unit cells that occur? Sketch them.

7. Aluminum has a density of 2.7 g/cm^3, an FCC crystal structure, and an atomic (molecular) weight of 26.98.
   a. What is the lattice parameter, in angstroms and in centimeters
   b. What is the diameter of an aluminum atom, in angstroms, if the atom is spherical in shape?

8. Sketch a cubic coordinate system, and show the following directions:
   a. [1,2,4]
   b. [1,−2,1]
   c. [3,2,−1]

9. Using any coordinate system, show the planes that have the following Miller Indices:

a. (1,2,3)
b. (2,−1,4)
c. (−2,4,3)
d. (0,1,2)

10. Sketch the (2,1,1) plane, and show whether the direction [0,−1,1] is in that plane.

11. Determine which phases are present in the following regions of the generic phase diagram:
   a. Region 1−2−3−4−1
   b. Region 17−18−20−19−18
   c. Region 13−14−6−13

12. Determine which phases are present and the compositions of those phases at the following temperatures and percentages of element B:
   a. 500°C and 40% B
   b. 2000°C and 90% B
   c. 1200°C and 75% B
   d. 2210°C and 85% B

13. Determine the amounts of the phases present at the temperatures and compositions of Problem 12.

14. Using the generic phase diagram (Fig. 2.3), write the equations for all eight invariant reactions, giving the temperature at which the reaction occurs, the proper phases in the reaction, and the name of the reaction.

15. Using the generic phase diagram (Fig. 2.3), create a cooling description for the following alloys:
   a. 68% B
   b. 73% B
   c. 35% B
   d. 25% B

16. Memorize the key temperatures and compositions of the Fe−C phase diagram so that you can sketch the diagram from memory.

17. Using the iron−iron carbide phase diagram of Fig. 2.6, describe the cooling of the following alloys:
   a. 3.0% C
   b. 1.3% C
   c. 0.3% C

18. Using the generic phase diagram (Fig. 2.3), consider an alloy of 95 percent B and 5 percent A which is cooling from 3400°C to room temperature to answer the following questions.

a. What is superheat of the alloy, in °C?
b. What is the freezing range of the alloy?
c. What are the invariant reactions and reaction temperature as the alloy cools from 3400° to room temperature?
d. What phases are present at 1000°C?
e. What is the composition of each of the phases present at 1000°C?
f. What is the amount of each of the phases present at 1000°C? What are the compositions of the microconstituents?
g. What microconstituents are present at 1000°C, and what are the compositions of the microconstituents a. 1100 b. 800 d. δ, η e. δ(57%B) η (98%B) g. Eutectic(70%B), η(97%B)

REFERENCES

1. Kalpakjian, S. Manufacturing Processes for Engineering Materials, 2nd ed., 1991, Addison-Wesley, Reading, MA, pp. 52–63.
2. Barrett, C. R., Nix, W. D., and Tetelman, A. S. The Principles of Engineering Materials, 1973, Prentice-Hall, Englewood Cliffs, NJ, pp. 36–64.

INTERNET SOURCES

Online course for material science: *http://vims.ncsu.edu/Contents/TOC.html*
Phase diagrams: *http://www.chem.umr.edu/info/diag/diag241.html*
The copper page: *http://www.copper.org/*
Aluminum: *http://www.aluminum.org*
Steel: *http://www.steel.org*
The World of the Microscope, from AT&T: *http://www.att.com/microscapes/*
Smart materials in airplanes: *http://www.eng.auburn.edu/department/ae/labinfo/ AAL/main.html*
Titanium: *http://www.titanium.org*
Zinc: *http://www.zinc.org*
Lead: *http://www.lead.org*

3
Mechanical Material Property Relationships

3.1 INTRODUCTION

The mechanical material properties most frequently used are yield strength, tensile strength, and the modulus of elasticity. These are obtained from the stress-strain diagram, and the strength values may be expressed as either engineering stress or true stress. The engineering stress-strain diagram is the basis for determining the yield strength, tensile strength and elastic modulus of materials, and these values are very important for design engineers. On the other hand, the true stress-strain diagram is needed to estimate the flow stress, which is used to determine the forces required to deform materials into the desired shapes via the various deformation processes.

The tensile test requires specially produced test specimens, which is often time-consuming; thus, hardness tests are frequently used to estimate the tensile strength values. The hardness test can be performed rapidly and is a type of compression test; thus, it is frequently used instead of the tensile strength test in manufacturing production lines as an indicator of tensile strength.

3.2 ENGINEERING STRESS-STRAIN

The engineering stress-strain diagram is important, in that the yield strength, tensile strength, and elastic modulus of a material can be clearly defined and obtained. From the design viewpoint, a high yield strength is generally desired, because it permits less material to transmit the load. On the other hand, from the viewpoint of deformation processes, plastic deformation does not start until the yield strength has been exceeded. Also, in deformation processing, the ultimate tensile strength is where thinning of the material, commonly called "necking," is thought to occur. Thus the designer prefers materials with high strengths, but the manufacturing engineers prefer materials with low yield strengths if deformation processing is to be used. The lower yield strengths would permit smaller presses, because lower forces would be needed to form the shape. The modulus of elasticity indicates the stiffness of the material; materials with a high modulus will deform less than materials with a low modulus for the same load.

The engineering stress-strain data for materials is based upon room temperature behavior; at elevated temperatures the yield and tensile strength values and the elastic modulus are much lower and the elongation values are much higher. The use of elevated temperatures in hot and warm deformation processing permits lower forces to be used to obtain the desired shape.

The engineering stress-strain curve is also very useful for obtaining the strain hardening exponent of a material. The maximum stress is easily shown on the engineering stress-strain diagram, and the strain at this stress is used to approximate the strain hardening exponent.

The basic relationships for engineering stress-strain terms follow.

Engineering stress:

$$S = P/A_o \qquad\qquad (3.1)$$

where

P = force applied
A_o = original area
S = engineering stress

Engineering strain:

$$e = \int dl/l_o \qquad\qquad (3.2)$$
$$= (l_f - l_o)/l_o \qquad\qquad (3.3)$$
$$= \delta l/l_o \qquad\qquad (3.4)$$

where

l = length
l_o = initial length
δl = change in length, $l - l_o$
e = engineering strain

An example of the traditional engineering stress-strain curve is indicated in Fig. 3.1. The data for this figure and for Fig. 3.2 is presented in Table 3.1. Since both engineering stress and engineering strain are divided

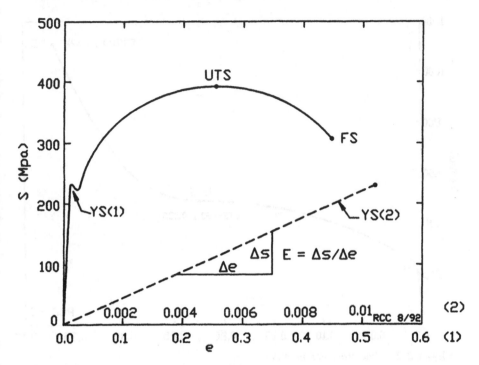

Figure 3.1 Engineering stress-strain diagram.

by constants relating to the original length and area, the same shape is obtained by a load-elongation curve (*P* versus δ*l*). Thus one advantage of the engineering stress-strain curve is that it is easily obtained from the load-elongation data. The slope of the engineering stress-strain curve is initially linear (for metals), because the stress increases to the yield stress. The slope of the curve is called the *modulus of elasticity* or *Young's modulus*. It can be expressed as:

$$E = \delta S/\delta e \qquad (3.5)$$

where

δ*S* = change in engineering stress (below yield point)
δ*e* = corresponding change in engineering strain for change in engineering stress

Another term, engineering strain rate (*ê*), is used when hot working occurs. It is defined as follows.

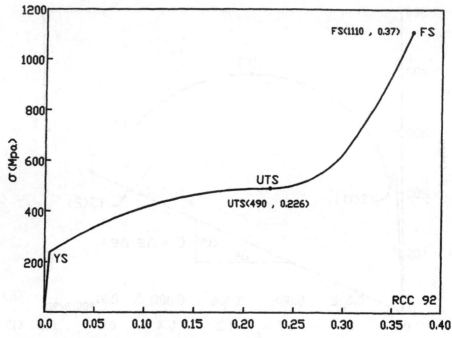

Figure 3.2   True stress-strain curve.

Engineering strain rate:

$$\dot{e} = de/dt$$

$$= \frac{d(\delta l/l_o)}{dt}$$

$$= \frac{d(l_f - l_o)/l_o}{dt}$$

$$= 1/l_o \times dl/dt \tag{3.6}$$

but

$$v = dl/dt \tag{3.7}$$

so

$$\dot{e} = v/l_o \tag{3.8}$$

where

$\dot{e}$ = engineering strain rate
e = engineering strain
t = time

Table 3.1 Data for Fig. 3.1 (Engineering Stress-Strain) and Fig. 3.2 (True Stress-Strain)

Load (kN)	Elongation (mm)	Area (m²)	Engineering stress (MPa)	Engineering strain	True stress (MPa)	True strain
—	—	0.000130	—	—	—	—
5	0.008		38.5	0.00016	38.5	0.00016
10	0.018		76.9	0.00036	76.9	0.00036
15	0.027		115	0.00054	115	0.00054
20	0.035		154	0.00070	154	0.00070
25	0.045		192	0.00090	192	0.00090
30	0.052		230	0.00104	230	0.00104
31	—		238	—	—	—
32.5	0.510		250	0.0102	253	0.0101
35	1.52		269	0.0304	277	0.0299
37.5	2.03		288	0.0406	300	0.0398
42.5	3.05		327	0.0610	347	0.0592
48	4.57		369	0.0914	403	0.0875
50	6.60		385	0.132	436	0.124
51	12.7	0.000104	392	0.254	490*	0.226
48	15.7	0.000085	369	0.314	565*	0.273
45.5	18.8	0.000059	350	0.376	771*	0.319
40	22.4	0.000036	307	0.448	1110*	0.370

*These true stress values are based upon actual specimen area.
The test specimen length initially is 50 mm.
Data is representative of a low-carbon steel.

l = length
l_o = original length
v = velocity (length/time)

The velocity is the velocity at which the load P is applied; that is, for a forging press it would be the velocity at which the dies are coming together. In mechanical testing, the velocity would be the velocity at which the load is applied that is pulling the test specimen.

3.3 TRUE STRESS-STRAIN

The primary difference between true stress-strain and engineering stress-strain is that the true values are based upon the actual (or current) values of length or area, whereas the engineering values are based upon the original (or starting) values of length or area. Since thinning or necking occurs at stresses above the true ultimate stress, it is desirable to keep the deformation stresses in the range between the yield stress and the ultimate stress. The relationships for true stress-strain for stress, strain, and strain rate follow.

True stress:

$$\sigma = P/A \tag{3.9}$$

where

σ = true stress
P = applied force
A = actual or instantaneous area

True strain:

$$\varepsilon = \int dl/l \tag{3.10}$$
$$= \ln (l/l_o) \tag{3.11}$$

where

ε = true strain
l = length
l_o = original length

An example of the true stress-strain curve is indicated in Fig. 3.2. Note that the true fracture stress is greater than the true tensile strength, whereas in the engineering stress-strain curve of Fig. 3.1, the engineering fracture stress is less than the engineering tensile strength.

True strain rate:

$$\dot{\varepsilon} = d\varepsilon/dt \tag{3.12}$$

$$= \frac{d(\ln l/l_o)}{dt}$$

$$= \frac{1}{l/l_o} \times l/l_o \times dl/dt$$

$$= 1/l \times v$$

$$= v/l \tag{3.13}$$

where

$\dot{\varepsilon}$ = true strain rate
ε = true strain
v = velocity
l = length

The relations for true strain rate and engineering strain rate are similar; the difference is that the engineering strain rate is based upon the initial or original length, whereas the true strain rate is based upon the actual or final length.

3.4 RELATIONSHIPS BETWEEN ENGINEERING STRESS-STRAIN AND TRUE STRESS-STRAIN

The values for true stress and true strain can be obtained from the engineering stress and engineering strain values when the stress is at or lower than the tensile strength (ultimate tensile strength). The specimen does not neck at stresses at or below the material tensile strength, and the elongation and contraction can be assumed to be uniform throughout the specimen rather than localized. The uniform elongation leads to the assumption of constant volume; that is:

$$V = A \times l = A_o \times l_o \tag{3.14}$$

where

V = volume
A = cross-sectional area
l = specimen length
A_o = original cross-sectional area
l_o = original specimen length

From this relation, it can be shown that:

$$l/l_o = A_o/A \tag{3.15}$$

Since

$$e = (l - l_o)/l_o \tag{3.16}$$

$$= l/l_o - 1 \tag{3.17}$$

then

$$l/l_o = e + 1 \tag{3.18}$$

and

$$A_o/A = e + 1 \tag{3.19}$$

Now

$$\sigma = P/A \tag{3.20}$$

$$= P/A_o \times A_o/A$$

$$= S \times (e + 1) \tag{3.21}$$

Thus Eq. (3.21) permits the determination of the true stress from the engineering stress and engineering strain. This relationship is valid only to the tensile strength (ultimate tensile strength) value and is not applicable in the range from the tensile stress to the fracture stress. True stress values above the tensile stress must be calculated from the cross-sectional areas in the necked region. Similarly, the true strain can be determined by:

$$\varepsilon = \ln (l/l_o) \tag{3.22}$$

$$= \ln (e + 1) \tag{3.23}$$

Equation (3.23) is also valid for strains only up to the tensile strength strain and is not valid for strains in the range from the tensile stress to the fracture stress. True strain values can be calculated from the actual elongation values.

One of the major advantages of true strain calculations is that true strain is additive, whereas engineering strain is not. What this implies is that:

$$\varepsilon_{13} = \varepsilon_{12} + \varepsilon_{23} \tag{3.24}$$

Thus true strain is mathematically consistent, which is very important in deformation processing, where large strains are frequently incurred. For most considerations in processing, true stress and true strain data is preferred over engineering stress and engineering strain data.

The slope of the true stress-strain curve in the elastic region (up to the yield point) can be used to determine Young's modulus, or the modulus of elasticity, when the amount of strain is small. The restriction to small amounts of strain generally holds for metallic materials. An equation for calculating the modulus of elasticity based upon true values would be:

$$E = \delta\sigma/\delta\varepsilon \qquad\qquad (3.25)$$

where

E = modulus of elasticity (Young's modulus)
$\delta\sigma$ = change in true stress, but below yield point and in the linear portion of the true stress-strain curve
$\delta\varepsilon$ = corresponding change in true strain

Since the strain levels are small, either true stress and strain values (Eq. 3.25) or engineering stress and strain values (Eq. 3.5) can be used, but traditionally the engineering values have been used.

3.5 STRAIN HARDENING RELATIONSHIPS AND STRESS-STRAIN DATA

For strain hardening materials, the general relationship relating true stress and true strain is applicable in the plastic region from the yield stress to the ultimate tensile stress. In this region the straining of the material increases the strength of the material. The relationship used is:

$$\sigma = K\varepsilon^n \qquad\qquad (3.26)$$

where

σ = true stress
K = strain hardening constant
n = strain hardening exponent
ε = true strain

From the engineering stress-strain diagram, the maximum load and ultimate tensile strength are easily observed, and one can readily obtain the engineering strain corresponding to the ultimate tensile strength. If the material follows Eq. (3.26), it can be shown that the numerical value of strain hardening exponent is equal to the true strain value at maximum load. Since the maximum load occurs at the ultimate tensile strength, the value of the engineering strain can be determined from the engineering stress-strain diagram, and the true strain can be found from Eq. (3.24):

$$\varepsilon = \ln (1 + e)$$

and

$$n = \varepsilon \text{ (at maximum load)} \tag{3.27}$$

From an examination of Eq. (3.26), a plot of the true stress and true strain data between the yield strength and ultimate tensile strength on logarithmic or log-log paper permit the determination of both n and K. The curve should be linear on log-log paper; the slope of the curve is n, and the intercept will give the value K. The intercept is the value of the true stress when the true strain has a value of 1.0, which may require extrapolation of the data. The slope value of n is the better estimate, but the value from Eq. (3.27) is another method for estimating the strain hardening coefficient.

3.6 HARDNESS RELATIONSHIPS FOR APPROXIMATING TENSILE STRENGTH

The tensile strength test is a destructive test in which the test sample is loaded until it fractures. The test usually requires machined samples and expensive testing equipment. For ductile materials, it has been shown (1) that true stress-strain curves in tension and compression are identical, so a compression test can be used to estimate a tensile strength.

The Brinell hardness test has successfully been used to estimate tensile strength values for ductile materials. The test applies a load to a 10-mm-diameter ball, and the diameter of the indentation is measured. This diameter and the load applied are converted to the stress applied to the curved surface area indentation. The value calculated, called the *Brinell hardness number* (*BHN*), is the stress applied; the units are kg/mm². The pressure multiplying factor (Q) for this type of loading is approximately 3, and the conversion factor of 9.8 converts the load in kg/mm² to N/mm² of megapascals (MPa). Thus the relation to convert BHN to ultimate tensile strength (UTS) values is obtained from:

$$\text{UTS (MPa)} = 1/Q \times \text{BHN} \times 9.8 \tag{3.28}$$
$$= 3.3 \times \text{BHN} \tag{3.29}$$

where

Q = pressure multiplying factor = 3
BHN = Brinell hardness number, in kg/mm²
UTS = ultimate tensile strength, in MPa
9.8 = conversion factor to convert kg force to N

The corresponding expression for the ultimate tensile strength in kpsi when the BHN number is used is:

$$\text{UTS (kpsi)} = 1/Q \times \text{BHN} \times 1.419 \qquad (3.30)$$
$$= 0.473 \times \text{BHN} \qquad (3.31)$$

where

Q = pressure multiplying factor = 3
BHN = Brinell hardness number, in kg/mm^2
UTS = ultimate tensile strength, in kpsi
1.419 = conversion factor to convert kg/mm^2 to kpsi

Since Eq. (3.31) is only an approximation for the ultimate tensile strength, it is often written in the rounded form as:

$$\text{UTS (kpsi)} = 0.500 \times \text{BHN} \qquad (3.32)$$

or

$$\text{UTS (psi)} = 500 \times \text{BHN} \qquad (3.33)$$

3.7 MATERIAL PROPERTIES AND MICROSTRUCTURE

The material properties are more closely related to the microstructure of the material than to the composition of the material. For low-carbon steels, the strength is related to the amount of the pearlite microstructure (which is the eutectoid structure); in aluminum alloys, the strength is related to the amount of the eutectic structure. The strengths are related to the volume amount of the phases present, rather than the weight fractions, which are typically calculated by the lever rule. For steels, however, the density of pearlite and ferrite are similar, so the volume fraction is also approximately the same as the weight fraction, and the weight fraction can be used as an estimate of the volume fraction.

The tensile strength (TS) of steel can be related to its Brinell hardness number for steels with pearlite (eutectoid) structures by:

$$\text{TS (psi)} = 500 \times \text{BHN}$$

or

$$\text{TS (MPa)} = 3.3 \times \text{BHN}$$

(Note that these equations do not give the same value, because the 500 is a rounded value and the 3.3 would need to be increased to 3.45 to give similar values.)

The tensile strength of pearlitic steels can be predicted from the amounts of the microconstituents and the type of pearlite. The strength values for the microconstituents are presented in Table 3.2.

Table 3.2 Tensile Strength Values of Microconstituents for
Carbon Contents Below 0.5 Percent

| | Tensile strength ||
Microconstituent	kpsi	MPa
Ferrite	80	550
Pearlite		
coarse	240	1700
medium	280	1900
fine	380	2600

Data from Ref. 2.

The amount of the microconstituents can also be found from the phase diagram in Chapter 2. Pearlite, at 0.8 percent carbon, is one of the micro-constituents; the other is ferrite, which has 0.025 percent carbon when formed and which can be considered to be zero for calculation purposes. Thus if a steel has 0.30 percent carbon and the microstructure is medium pearlite, the tensile strength can be estimated by finding the amounts of pearlite and ferrite and then calculating the strength. Note that there is a large variation in the properties of the pearlite as a function of the platelet size; fine pearlite platelets result from fast cooling, whereas coarse pearlite platelets are a result of slow cooling rates. Since the density of pearlite and ferrite are similar, the volume percent and weight percent values can be considered equivalent. The phase diagram permits calculation of the weight percent values, but the volume percent values are needed to calculate the strength of the mixture. As previously mentioned, for low-carbon steels the volume fraction is approximately equivalent to the weight fraction and thus the lever rule can be used. This assumption does *not* work for most other materials. Thus, using the lever rule, the amounts of pearlite and ferrite can be found as:

$$\% \text{ pearlite} = (0.30 - 0)/(0.80 - 0) \times 100 = 37.5\% \ (0.375 \text{ decimal})$$
$$\% \text{ ferrite} = (0.80 - 0.30)/(0.80 - 0) \times 100 = 62.5\% \ (0.625 \text{ decimal})$$

The tensile strength of the steel alloy, from the values in Table 3.2, would be:

$$TS = 0.375 \times 280 + 0.625 \times 80$$
$$= 155 \text{ kpsi} = 155,000 \text{ psi}$$

or

$$TS = 0.375 \times 1900 + 0.625 \times 550$$
$$= 1060 \text{ MPa}$$

The general rule for determining the properties of materials with two phases, written for the determination tensile strength, is:

$$\sigma = \sigma_1 \times f_1 + \sigma_2 \times f_2 \qquad (3.34)$$

where

σ = strength of mixture
σ_1 = strength of microconstituent 1
σ_2 = strength of microconstituent 2
f_1 = volume fraction of microconstituent 1
f_2 = volume fraction of microconstituent 2

Other material properties that are a function of the volume fraction of the material can also use the form of Eq. (3.34). Thus, Eq. (3.34) is not limited to steels, but can be used for composites and other two-phase materials. The key step is to remember to obtain volume fractions and not to use mass or weight fractions. With the phase diagram of Fig. 3.3 and the data of Table 3.3, the calculations require the determination of the volume fraction from the mass fractions.

The volume fractions can be determined from the mass fractions. For a two-phase system, the relationship for phase 1 is:

$$fv(1) = \frac{fm(1)/d1}{fm(1)/d1 + fm(2)/d2} \qquad (3.35)$$

where

$fv(1)$ = volume fraction of phase 1
$fm(1)$ = mass or weight fraction of phase 1
$fm(2)$ = mass or weight fraction of phase 2
$d1$ = density of phase 1
$d2$ = density of phase 2

If one has an alloy of 4 weight percent aluminum, then from the phase diagram data of Fig. 3.3 the mass or weight fractions are:

$$fm(\text{aluminum}) = (12.6 - 4.0)/(12.6 - 1.65) \times 100 = 78.54\% \ (0.79)$$
$$fm(\text{eutectic}) = (4.0 - 1.65)/(12.6 - 1.65) \times 100 = 21.46\% \ (0.21)$$

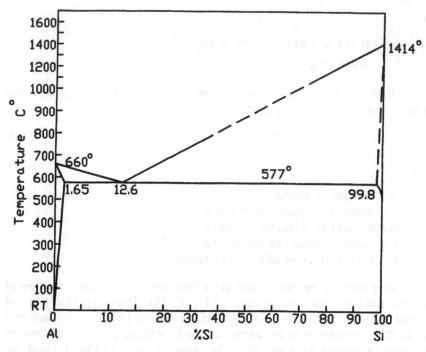

Figure 3.3 Aluminum-silicon phase diagram. (Developed from data from Ref. 3.)

The volume fractions for these weight fractions are thus:

$$fv(\text{aluminum}) = \frac{0.79/2.7}{0.79/2.7 + 0.21/3.2}$$
$$= 0.817 = 0.82$$
$$fv(\text{eutectic}) = \frac{0.21/3.2}{0.79/2.7 + 0.21/3.2}$$
$$= 0.183 = 0.18$$

Table 3.3 Hypothetical Data for Al-Si Eutectic System

Property	Primary Al	Eutectic
Density (g/cc)	2.7	3.2
Strength (kpsi)	10	20

The tensile strength of the alloy would then be:

$$\sigma = 0.82 \times 10 + 0.18 \times 20$$
$$= 11.8 \text{ kpsi}$$

The calculation of volume fractions is utilized again in Chapter 11 on powder metallurgy and powder processing. Volume fraction is related to the mechanical properties, whereas mass fraction is used for determining costs and mix ratios; these are illustrated in Chapter 11.

3.8 SUMMARY

The key mechanical material property used by designers is the yield strength of a material. The ultimate tensile strength is a property needed by manufacturing engineers to estimate the forces required to shape the material by deformation processing. The engineering stress-strain diagram clearly indicates the yield point and ultimate tensile strength, and provides a method for estimating the strain hardening coefficient. The modulus of elasticity is the slope of the engineering stress-strain curve in the elastic region.

The true stress-strain approach is the best method to determine the actual stress occurring in a material under loading. It is the correct approach to determine the amounts of deformation, for true strains are additive whereas engineering strains are not additive.

The hardness test can be used to estimate the engineering tensile strength, and frequently is because it is a faster and lower-cost test.

Material properties are predicted better by the amounts of the microconstituents than by composition analysis. The material properties are a function of the volume fraction of the microconstituents rather than of the weight fraction; this is very important for most aluminum alloys. Weight fraction, however, is useful for determining costs and mix ratios.

3.9 EVALUATIVE QUESTIONS

1. Using the data in Table 3.1, make an engineering stress-strain diagram. Compare the results with Fig. 3.1.

2. Using the data in Table 3.2, make a true stress-strain diagram. Compare the results with Fig. 3.2.

3. Using the elongation data for the loads of 35, 42.5, and 48 kN from Table 3.1, verify that the true strains are additive and that the engineering strains are not additive.

4. Plot the appropriate true stress-strain data of Table 3.1 on log-log paper or on the computer and obtain the strain hardening exponent and the strain hardening constant. Also, use Eq. (3.27) to estimate the strain hardening exponent. Discuss the difference.

5. Estimate the tensile strength in both unit systems, kpsi and MPa, for a material with a Brinell hardness number of 180.

6. The following load-elongation data was obtained from a sample 50 mm in length and 12.8 mm in diameter (kN = kilonewtons and 1 MPa = 1 N/mm^2)

	Load (kN)	Elongation (mm)
Start	0	0.0
	10	0.010
	20	0.020
Yield	25	0.025
	30	11.2
Max. load	38	18.7
Fracture	18	24.5

   a. What is the engineering ultimate tensile strength (MPa)?
   b. What is the true ultimate tensile strength (MPa)?
   c. Estimate the strain hardening coefficient (n)?
   d. Estimate the strain hardening constant (K) in the strain hardening equation, Eq. (3.26).
   a. 295 MPa; b. 406 MPa; c. 0.317; d. 584 MPa

7. Using the iron−iron carbide phase diagram in Chapter 2 (Fig. 2.6) and the data of Table 3.2, estimate the amount of ferrite and pearlite, and the tensile strength of the alloy, in MPa, for a pearlite of fine structure and a carbon content of 0.25 percent. Also, estimate the BHN for this material.

8. Using the aluminum-silicon phase diagram (Fig. 3.3), estimate the tensile strength of the alloy for a silicon content of 8 percent.

9. Using the iron−iron carbide phase diagram of Fig. 2.6, estimate the microconstituents present at 600°C for the following compositions:
   a. 0.3 percent C
   b. 0.8 percent C
   c. 1.2 percent C

REFERENCES

1. Kalpakjian, S. Manufacturing Processes for Engineering Materials, 2nd ed., 1991, Addison-Wesley, Reading, MA, pp. 52–63.
2. Ludema, K. C., R. M. Caddell, and A. G. Atkins. Manufacturing Engineering— Economics and Processes, Prentice Hall, Englewood Cliffs, NJ, 1987, pp. 126– 128.
3. Metals Handbook, Volume 8: Metallography, Structures, and Phase Diagrams, 8th ed., American Society for Metals, 1973, p. 263.

INTERNET SOURCES

Stress-strain, hardness: *http://www.tiniusolsen.com/tech.html*
Yahoo material science links: *http://www.yahoo.com/Science/Engineering/Material_ Science/*
Thermal properties: *http://www.mayahtt.com/tlab/props.htm*

4
Methods for Increasing Mechanical Material Properties

4.1 INTRODUCTION

There is a great need to increase the mechanical properties of materials above their base level for a material. Increased properties lead to smaller section sizes, lower weights, lower energy consumption for manufacturing, and usually lower costs. The lower-weight, or "diet," materials make possible new products and new designs not previously possible, such as in spacecraft and aircraft structures, lighter-weight and higher-performance automobiles, and portable computers, among many other new products. The focus of this chapter is on metallic materials; some of the methods presented here may *not* be appropriate for other materials.

There are four methods commonly used to increase the strength of metallic materials:

1. Solution hardening (alloying)
2. Grain size control
3. Strain hardening (cold work)
4. Heat treatment

The first two methods give moderate increases in yield strength, in the range of 5–30 percent, whereas the last two methods give much larger increases in strength, such as 50–300 percent increase in yield strength. Each of the methods will be presented in detail, with an emphasis on the last two methods.

Deformation generally takes place by means of the mechanism called *slip*, whereby a plane of atoms will slide across another plane of atoms. This slip mechanism occurs within a grain, and techniques that increase the strength of the material prevent or hinder the slip mechanism. A description of slip planes and slip directions using Miller indices is presented in Chapter 2. There are different methods for preventing slip, and the four methods of increasing strength are different approaches to preventing slip from occurring.

4.2 SOLID SOLUTION HARDENING

In solid solution hardening, or *alloying*, a different atom is added to the material. A pure metal or compound has a specific unit cell and crystal structure. Since all of the elements have different atom sizes, when a second atom is added, the unit cell will become distorted and slip will be made more difficult. The second atom may substitute for the first atom at a lattice site; or if it is much smaller, it may be an interstitial atom and locate itself in one of the void spaces in the unit cell.

If the atom is larger or is similar in size, it generally substitutes for the atom at one of the lattice points. If it is larger, it will distort the surrounding unit cells and make slip more difficult. If the atom is smaller in size, the adjacent unit cells will tend to be slightly distorted because of the smaller atom.

When the second atom is much smaller in size, it will tend to fill one of the void spaces (commonly called *holes*) in the lattice. The smaller atom, however, is not a perfect match because of differences in size, electrical properties, magnetic properties, and so on. These differences will cause a lattice strain and prevent slip from occurring.

The effects of solid solution strengthening vary approximately with the square root of the concentration of the second element (1). Thus the strength can be approximated by:

$$\sigma = \sigma_1 + K1 \times \sqrt{C} \tag{4.1}$$

where

σ = tensile strength
σ_1 = tensile strength constant for solution strengthening
$K1$ = solid solution concentration constant
C = solid solution concentration, atomic percent

Thus the effects of increases in strength for solution increases tend to be small, that is, at the 5–20 percent level.

4.3 STRENGTHENING BY GRAIN SIZE CONTROL

Grain boundaries prevent slip from occurring because two adjacent grains have their planes at different orientations. Thus if slip occurs along a plane in one atom, the change in orientation through the next grain makes slip more difficult. Materials with numerous grains reduce the amount of slip occurring, and the strength of a material increases as the grain size decreases.

The area of a grain is proportional to the square of the diameter of the grain. The strength of the material increases with decreasing grain area; therefore the strength can be inversely related to grain area or inversely to the square root of the grain diameter. The strength of material can be related to its grain size by (2):

$$\sigma = \sigma + K2/\sqrt{d} \qquad\qquad (4.2)$$

where

σ = tensile strength
σ_2 = typical strength of single large grain
$K2$ = grain boundary strengthening constant
d = grain size

This relationship indicates that strength varies inversely with the square root of the grain size, and thus the effects of increasing strength by grain size is limited to a maximum of approximately 30 percent; the typical increases are 5–15 percent.

4.4 STRENGTHENING BY STRAIN HARDENING

In cold working, or strain hardening, line defects called *dislocations* are created that make it more difficult for slip to occur. The more dislocations, the stronger the material. If too many dislocations are generated, the material may crack; thus there is a maximum limit to the amount of cold work for a material. Also, since more cold work occurs at the surface than at the center of the material, the strength properties of the material will be higher at the surface than in the center. In general, it is desirable to have the higher properties at the surface because of wear and because high stresses tend to be at the surface. For fatigue applications, however, compressive rather than tensile stresses are desired. The removal of the strain hardening effects can be accomplished by recrystallization of the material. During recrystallization, the dislocations in the material move to the grain boundaries or move to the surface and leave the material in a "strain-free" condition.

4.4.1 Hot, Cold, and Warm Working

Hot, cold, and warm working are related to the recrystallization temperature of a material and to room temperature. The recrystallization temperature is a function of the absolute melting temperature of the material and the amount of cold work, or strain hardening, that the material has undergone.

Thus it is necessary to discuss cold work before discussing recrystallization and the recrystallization temperature.

Cold work, or strain hardening, occurs when the material is plastically deformed and dislocations are created. As the material is cold worked, the yield strength, tensile strength, and hardness increase, whereas the ductility of the material decreases. The amounts of increase can approach 100 percent (50–150 percent range)—that is, the yield strength can double—and the increase in tensile strength is usually somewhat less, though still very large. During the process of strain hardening, the energy imparted to the material and the dislocations generated make the material brittle. The dislocations and brittleness can be removed by heating the material to relieve the strains caused by deformation and the formation of new strain-free grains. The temperature at which the new grains form is called the *recrystallization temperature*.

The recrystallization temperature is defined as the temperature at which the recrystallization is 95 percent complete within one hour of heating at the specified temperature. The recrystallization temperature for a material is approximately one-third to one-half of its melting temperature when using an absolute temperature scale. A minimum amount of strain hardening must occur, at least 5 percent, before any recrystallization can occur. The percent of cold work is often expressed as a percent reduction in thickness of the material during rolling, or as a percent reduction in cross-sectional area when the width can also change. Thus, the expression for cold work is:

$$\%CW = (A_o - A_f)/A_o \times 100 \tag{4.3}$$

where

$\%CW$ = percent cold work
A_o = original cross-sectional area
A_f = final cross-sectional area

For rolled materials, where the cross-sectional width does not change, the expression can be reduced to:

$$\%CW = (t_o - t_f)/t_o \times 100 \tag{4.4}$$

where

$\%CW$ = percent cold work
t_o = original thickness
t_f = final thickness

The more cold work a material has, the lower the recrystallization temperature. This is because recrystallization is a function of the total energy, that is, the strain energy and the thermal energy. Thus if there is more strain energy (i.e., more cold work), less thermal energy will be required. Note that the expression (Eq. 4.4) for cold work in rolled materials is similar to that for engineering strain (Eq. 3.3) in compression loading.

Cold working is defined as the working of a material below its recrystallization temperature; hot working is the working of a material above its recrystallization temperature. When a material is worked above its recrystallization temperature, the material recrystallizes immediately as the working occurs, so no strain hardening builds up in the material. The advantage of hot working is that less forces are needed to deform the material; but the major disadvantages are that the high temperatures may oxidize the surface of the material and that dimensions and tolerances on the material are harder to control.

Warm working is a cold working process in that it is done below the recrystallization temperature; but it is done above room temperature, and thus the forces required to form the shape are reduced. For cold-worked materials, annealing is used to remove the effects of cold work and possible cracking if further working is needed. When metals must undergo severe plastic deformation, this is often done in stages, with intermediate anneals (recrystallization operation) between the stages. Materials with low melting temperatures and low recrystallization temperatures, such as lead, cannot be cold worked or warm worked at room temperature, for room temperature is above the recrystallization temperature for these materials.

The large increases in yield strength and tensile strength as a function of the amount of cold work is indicated in Table 4.1. Hardness increases are similar to the increases in tensile strength for the aluminum and copper alloys illustrated. It must also be noted that the ductility, as determined by the percent elongation, decreases with increasing amounts of cold work.

The influence of cold work is very large upon the mechanical properties of the material. For fully strain-hardened materials, the yield strengths increase from 100 to 500 percent, and the tensile strengths increase from 50 to 100 percent. Hot working is often performed in the preliminary forming stages, to obtain the near net shape; cold work is done for the final stage, to obtain the desired mechanical properties, surface finish, and tolerances.

Two major problems that frequently occur with cold working is that the strength increases are not uniform throughout the material and that the changes vary with respect to direction and depth. The surface is worked more than the inside of the material, so the surface has higher strength values. The differences in the properties with respect to direction can be evaluated by means of the strain ratio and two anisotropic coefficients.

Table 4.1 Tensile Strength, Yield Strength, Hardness, and Ductility Values* for Various Materials as a Function of Cold Work

Material condition	%CW	Tensile strength (kpsi)	Yield strength (kpsi)	Hardness (BHN)	Ductility (% elongation)
3003 Al (aluminum)					
O (annealed)	—	16	6	28	40
H12	18	19	18	35	20
H14	35	22	21	40	16
H16	55	26	25	47	14
H18	75	29	27	55	10
5052 Al					
O (annealed)	—	28	13	47	30
H32	18	33	28	60	18
H34	35	38	31	68	14
H36	55	40	35	73	10
H38	75	42	37	77	8
C10400 (copper)					
O (annealed)	—	34	11	10R (B)**	45
H01	11	38	30	25	30
H02	21	42	36	40	14
H04	37	50	45	50	6
H08	61	55	50	60	4
H10	69	57	53	62	4

*Reprinted with permission from ASM International (Ref. 2).
**Rockwell B Hardness values for C10400.

4.4.2 Strain Ratio and Anisotropic Coefficients

In the evaluation of materials for deformation, the strain ratio and anisotropic coefficients have been useful to predict the behavior of sheet metal materials. Generally, deformation processes are classified into bulk deformation, such as forging, where there are large changes in all dimensions, and sheet metal working, where the work is basically two-dimensional and little change occurs in the thickness direction. The strain ratio and anisotropic coefficients have been used by the automotive industry to evaluate materials and processes for forming the various sheet metal products, such as doors, fenders, roofs, hoods, and trunk lids.

In sheet metal products, the mechanical properties can vary directionally. The reference direction is the direction of rolling; other commonly used directions are 45° and 90° from the rolling direction. Figure 4.1 illustrates

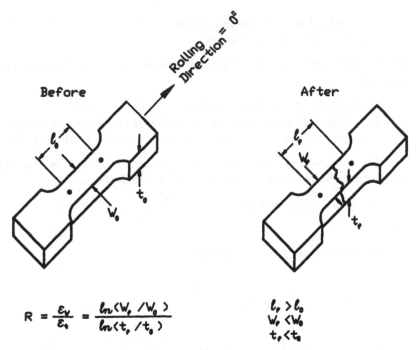

$$R = \frac{\varepsilon_v}{\varepsilon_t} = \frac{\ln (W_f / W_o)}{\ln (t_f / t_o)}$$

$l_f > l_o$
$W_f < W_o$
$t_f < t_o$

R_o = Values when load applied in Rolling Direction as Indicated.
R_{45} = Values when load applied on sample made at
         45° angle to Rolling Direction.
R_{90} = Value when Load applied on sample made at
         90° angle to Rolling Direction.

Figure 4.1 Sample for strain ratio determination in sheet metal.

the sample used for evaluating the true width and thickness strains when calculating the strain ratio. The expressions are:

$\varepsilon_w = \ln (w_f/w_o)$
$\varepsilon_t = \ln (t_f/t_o)$
$R = \varepsilon_w/\varepsilon_t$

where

   R = strain ratio
   ε_w = width strain
   ε_t = thickness strain
   w_o = original width before strain
   w_f = final width after strain

t_o = original thickness before strain

t_f = final thickness after strain

The R values can be calculated for the various directions, which are designated as:

R_0 strain ratio in rolling direction

R_{45} strain ratio at 45° to rolling direction on rolling surface plane

R_{90} strain ratio at 90° to rolling direction on rolling surface plane

These strain ratios can now be used to determine various anisotropic coefficients. The two most commonly used coefficients are:

$$R_m = (R_0 + 2R_{45} + R_{90})/4$$

where

R_m = mean anisotropic coefficient

and

$$R_p = (R_0 - 2R_{45} + R_{90})/2$$

where

R_p = planar anisotropic coefficient

If the values of the strain ratio in the three directions (0°, 45°, and 90°) are all equal to 1.0, the material is said to be isotropic. This would also cause the mean anisotropic coefficient to have a value of 1.0, and the planar anisotropic coefficient would be zero. If the values of the strain ratio in the three directions are all equal to a value other than 1.0, the material is said to have planar isotropy and the planar anisotropic coefficient will be zero, but the mean anisotropic coefficient will be a value other than 1 but equal to the individual strain ratio value.

It is desirable to have a high value for the mean anisotropic coefficient, becauses it implies a good resistance to thinning and a good thickness strength. It is desirable to have zero value for the planar anisotropic coefficient, because the larger the value, the greater the tendency for earing during deep drawing. Typical strain ratio values and/or anisotropic coefficients for typical materials are presented in Table 4.2. Note that the material structure also has an influence upon the strain ratio and the anisotropic values. In general, FCC materials have lower values for the mean anisotropic coefficient than do BCC materials. Hexagonal close-packed materials with a high c/a ratio, such as zinc, have low values of the mean anisotropic coefficient, whereas HCP materials with a low c/a ratio, such as titanium, have the highest mean anisotropic coefficients.

Table 4.2 Strain Ratios and Anisotropic Coefficients for Typical
Engineering Materials

Material and structure		R_0	R_{45}	R_{90}	R_m	R_p
Normalized steel	BCC	0.9	1.1	0.9	1.0	−0.20
Copper, brass	FCC				0.5−0.9	
Zinc	HCP				0.2	
Lead	FCC				0.2	
Titanium	HCP				3.0−6.0	
Killed steel (Al)	BCC	1.6	1.4	1.9	1.6	+0.35
Low-N, -C steels	BCC				1.8−2.4	
Annealed steel	BCC	1.3	1.0	1.4	1.2	+0.35
HSLA*—hot rolled	BCC		·		0.8−1.0	
HSLA*—cold rolled	BCC				1.0−1.4	

*HSLA = High-strength low-alloy
Adapted from Ref. 3, 4, and 5.

In summary, a high mean anisotropic coefficient gives a high resistance
to the thinning of materials, which can occur in sheet metal working operations such as deep drawing and bending. On the other hand, a low planar
anisotropic coefficient reduces the tendency for earing, or uneven surfaces,
in deep drawing. Control of the anisotropic coefficients is important, as is
the increase in strength in strain hardening operations.

4.5 HEAT TREATMENT

4.5.1 Terminology

Heat treatment is the heating and cooling of alloys in the solid state to obtain
the desired mechanical properties. Significant increases, such as 100 percent
or more, can be obtained in the tensile strength of a steel by heat treatment.
Since the design of products frequently depends upon material properties,
the use of heat-treated materials can reduce the size of a component needed
to accomplish the same design function.

There are two general types of hardening by heat treatment:

1. Allotropic transformation suppression (martensite formation)
2. Precipitation (dispersion) hardening

The allotropic transformation suppression method, more commonly
called *martensite formation hardening*, requires that the alloy undergo an
allotropic transformation. An allotropic transformation occurs when the alloy

changes its crystal structure upon cooling. The most notable metal that undergoes an allotropic transformation is iron, but some alloys of titanium and some alloys of manganese can also undergo the martensite transformation. Iron changes its crystal structure from FCC (austenite) to BCC (ferrite) during cooling. Under quenching or rapid cooling conditions, this transformation is suppressed and the austenite (FCC iron) transforms to martensite (body-centered tetragonal iron) via a diffusionless shear mechanism instead of transforming to ferrite (BCC iron) by nucleation and growth. Since martensite is produced by rapid quenching, it does not appear on the phase diagram (because it is not an equilibrium product, i.e., produced by slow cooling). Most alloys do not undergo an allotropic transformation and thus cannot be hardened by the allotropic transformation suppression method.

The precipitation hardening method requires the formation of a coherent precipitate from a supersaturated solid solution. At high temperatures, a metal will dissolve more of a second metal (alloy); but when the metal is quenched, the second metal is trapped and thus a supersaturated solid solution results. Most metals form coherent precipitates, but they form them only with certain metals and not all metals. For example, aluminum will form precipitational hardening alloys with copper or zinc but not with either manganese, silicon, or magnesium. However, if both silicon and magnesium are alloyed with aluminum, a coherent precipitate is formed and precipitation hardening can occur.

There are numerous terms used in heat treatment, and some of the more frequently used terms are defined. We group them here into three categories: (1) general terms, which apply to both methods of heat treatment; (2) ferrous terms, which apply to the allotropic transformation methods; and (3) precipitation hardening terms.

4.5.1.1 General Terms

Hardness: The resistance to penetration.

Anneal (full or conventional): A heat treatment process of heating and slow cooling used to soften materials.

Homogenization: A heating and holding at a temperature to produce a homogeneous composition. When a two-phase structure is heated to a high temperature where one of the phases is dissolved in the other, the atoms must diffuse to form a uniform composition at the atomic level.

Hardening: A process by which the hardness of a metal is increased. This usually implies hardening by heat treatment, but hardening can also be done by alloying or work hardening (strain hardening).

Microconstituent: A phase or combination of phases that appear as material structure under microscopic examination. Bainite and pearlite are considered microconstituents, whereas martensite, ferrite, cementite, and austenite are considered phases and also are considered microconstituents when present as separate grains. The combination of phases frequently is the result of a eutectic or eutectoid reaction where two phases nucleate simultaneously, and these are recognized as microconstituents. The pearlite microconstituent is the result of the eutectoid reaction in steel. Microconstituents have a major influence on the mechanical properties of a material.

Grains: Individual crystals in metals. A grain consists of numerous unit cells connected and aligned in the same orientation.

Supersaturated solid solution: A solid solution that has more solute dissolved in the solution than permitted by equilibrium cooling. It is unstable, and the excess solute will try to precipitate out; but the diffusion of the atoms is extremely slow and the structures are relatively stable. The application of heat will cause the precipitation process to increase rapidly. Martensite is an example of a supersaturated solid solution, but any alloy with a trapped second phase is a supersaturated solid solution.

4.5.1.2 Ferrous Terms (Steels and Cast Irons)

Figures 3.2 and 4.2 can be used to locate some of the phases and microstructures described.

Austenite: A solid solution of one or more elements (usually including carbon) in the face-centered cubic crystal structure form of iron (often called *gamma* iron).

Ferrite: A solid solution of one or more elements (usually including carbon) in the body-centered cubic crystal structure form of iron. There are two regions of ferrite on the phase diagram: The lower-temperature region (<910°C) designated *alpha* ferrite, is the one most referred to; the higher-temperature ferrite (1395°C–1535°C) is designated *delta* ferrite.

Carbide: A compound of carbon with one or more metallic elements. Iron carbide, or cementite, is a compound of iron and carbon that has the approximate chemical compound formula of Fe_3C and is characterized by an orthorhombic crystal structure. There are other iron carbides, but the most common is the Fe_3C form.

Cementite: See *Carbide*; the terms are often used for each other; however, cementite is a particular carbide (Fe_3C), whereas carbide is more general.

Pearlite: The lamellar aggregate of ferrite and cementite. It is formed at temperatures above those at which bainite or martensite are formed. The lamellar formed are very thin and resolvable only with microscopic equipment with magnifying powers of 50× or more. It is a microconstituent and is not a phase, but a mixture of two phases.

Martensite: An unstable constituent in quenched steel, formed without diffusion and only during cooling below a certain temperature known as the M_s, or martensite start, temperature. The structure is characterized by its acicular appearance on the surface of a polished and etched specimen. Martensite is the hardest of the transformation products of austenite. It is a supersaturated solid solution of carbon in iron having a body-centered tetragonal structure.

Bainite: The ferrite-cementite aggregate formed by the transformation of austenite at temperatures below those at which pearlite forms and above those at which martensite forms. The ferrite nucleates first, and the cementite forms as platelets inside the ferrite and not at the grain boundaries. The fine, thin plates make the bainite more ductile than pearlite, which has coarse cementite plates.

Hardenability: The ability or ease with which one can form martensite. Materials of low hardenability do not form martensite easily, whereas materials of high hardenability form martensite easily.

Eutectoid steel: A steel of the eutectoid composition. The eutectoid composition in a pure iron-carbon alloy is approximately 0.8 percent C, but variations are found in commercial steels and particularly in alloy steels, in which the carbon content of the eutectoid is slightly lower.

Isothermal transformation: The process of transforming austenite to ferrite/pearlite, pearlite, or bainite by quenching the austenite to the desired temperature and holding at temperature until the transformation of austenite to the desired decomposition product(s) is complete.

4.5.1.3 Precipitation Hardening Terms

Coherent precipitate: A region where solute atoms concentrate to the degree necessary to meet the composition of the second-phase precipitate, but no interface (grain boundary) exists between the parent and the second phase. This causes a large strain on the lattice and prevents dislocation movement.

Incoherent (noncoherent) precipitate: The second phase has its own crystal lattice, and an interface (grain boundary) exists between the parent and the second phase (precipitate). There is little strain on the lattice with the incoherent precipitate, and dislocation movement is not as difficult.

Artificial aging: Heating a material above room temperature to hasten the formation of the desired precipitate (coherent or incoherent) and to obtain a desired precipitate size range.

Natural aging: Allowing the precipitate to form by itself without heating the material above room temperature.

Overaging: Continuing the artificial aging process for longer times or at higher temperatures so the precipitate forms and grows to a size greater than the optimal precipitate size, with a corresponding reduction in mechanical properties.

4.5.2 Procedure for Hardening

The procedure for either hardening process (allotropic transformation suppression or precipitation hardening) can be generalized into the same three steps:

1. Solution heat treatment (form a homogeneous solid solution)
2. Quench (form a supersaturated solid solution)
3. Reheat

For the allotropic transformation suppression process, the first step is to heat the material into the austenite region and hold it long enough to obtain a uniform austenite composition (homogenization). The second step, quenching is to suppress the allotropic transformation of austenite to ferrite and to obtain martensite instead. The third step, reheating, is to temper the martensite to provide sufficient ductility as it is extremely brittle in the untempered condition.

For the precipitational hardenable materials, the first step is to heat the alloy from the two-phase region up into the single-phase solid region and hold it until the second phase is dissolved and uniformly dispersed in the single phase. The second step is to quench to form a supersaturated solid solution and to prevent the second phase from forming and growing into a coarse structure. The third step, reheating, is to form a widely dispersed fine precipitate of the second phase throughout the parent phase. When the precipitate is of the desired size, the material is cooled rapidly back to room temperature to prevent overaging.

4.5.3 Transformation Diagrams

The diagrams that describe the transformation of austenite into martensite, bainite, or pearlite are frequently called TTT diagrams. The TTT stands for *time-temperature transformation*, and the diagram is a temperature-versus-time plot indicating the transformation of austenite. The temperature scale is related to that on the equilibrium iron–iron carbide phase diagram. Figure

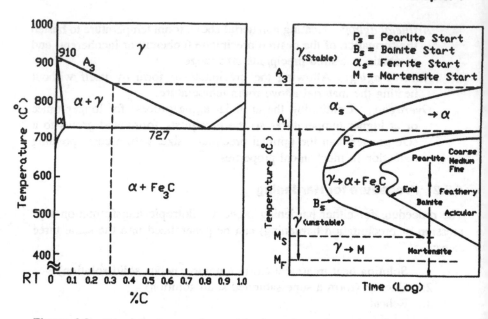

Figure 4.2 Phase diagram section and isothermal transformation diagram.

4.2 presents the pertinent part of the phase diagram and the TTT diagram for a hypoeutectoid steel. Note that the martensite start and martensite finish temperatures are not on the equilibrium phase diagram, since martensite is not an equilibrium structure. The different structures—martensite, bainite, and pearlite—have quite different properties with respect to strength, hardness, and ductility. Martensite has the highest hardness and strength, but poor ductility; bainite has high strength, hardness, and ductility; fine pearlite has moderate strength, hardness, and ductility; and coarse pearlite has poor strength, hardness, and ductility. The ductility of martensite can be improved by tempering (reheating), but the hardness and strength values are reduced slightly.

The transformation diagrams are very dependent upon the material composition, and some alloying elements for steel, such as chromium, molybdenum, nickel, and manganese, will delay the time at which the transformation of austenite starts and thus make it easier to form martensite. These elements will also effect the martensite start and finish temperatures. For steels with very low alloy contents and carbon levels below 0.10 percent, it is difficult to form martensite, because we cannot cool (quench) the steel fast enough. The hardness of the martensite formed is also dependent upon the carbon content, and steels with higher carbon levels, such as 0.4–0.6

percent, will be much harder, stronger, and less ductile than steels whose carbon levels are 0.1–0.2 percent. One of the problems of the quenching operation is that distortion (warping) of the part may occur; in very hard materials, cracking may occur if the tempering is not started immediately.

4.5.4 Age-Hardenable Materials

The heat treatment of age-hardenable materials follows the same procedure as that of the allotropic transformation suppression hardening, but the purpose of the last step is different. Figure 4.3 shows a section of a phase diagram and the microstructures obtained. The first step in the procedure is to heat a two-phase structure (structure 0) into a single-phase region, as indicated by temperature 1 and microstructure 1 of Fig. 4.3. If the structure is slowly cooled from temperature 1 to room temperature, the original coarse structure, as indicated by microstructure 0, will be obtained again. If, as the procedure specifies, the specimen is quenched, a supersaturated solid solution will be obtained, as indicated by microstructure 2. The third step is the reheating of the quenched structure to obtain the optimally aged structure, as indicated by microstructure 3. This is when the precipitate is just begin-

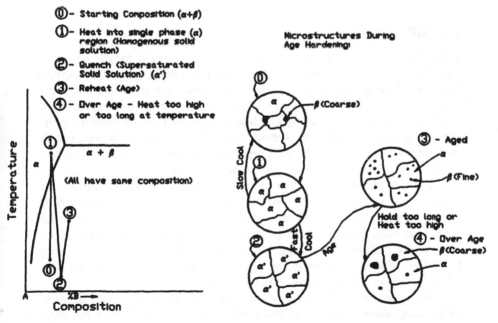

Figure 4.3 Phase diagram indicates temperature variation and microstructures for age hardening materials.

ning to form. One of the requirements is that a coherent second phase must be formed; otherwise the material is not age hardenable. If the material is heated at too high a temperature or for too long a time period, the overaged structure indicated by microstructure 4 will be obtained. The strength and ductility values will be lower than that of an optimally aged structure.

Aluminum and copper alloys are frequently age hardened, since they do not undergo allotropic transformations. The aluminum and copper alloys that cannot be age hardened may be hardened by strain hardening, commonly called *work hardening*. Some steels are precipitation hardened if the alloy contents are so high that the alloy does not undergo an allotropic transformation.

Dispersion hardening is a more general term than age hardening: The second phase is finely dispersed throughout the primary phase by means other than precipitation from a supersaturated solid solution. This can be done by mechanically mixing the second phase and using powder metallurgy processes, internal oxidation of particles, or other means. Carbide tools are formed by the powder metallurgy process, and this is dispersion hardening rather than age hardening; but the structure is similar to that of an age-hardened structure.

4.6 SUMMARY

Material properties comprise a critical parameter needed for the design of manufactured products. There are four methods generally used to increase the material properties of metals: alloying, grain size control, strain hardening, and heat treatment. Substantial increases in strength and hardness can be obtained by strain hardening and heat treatment. Thus the material properties vary not only with the specific material, but also with the specific hardening treatment. The different properties provide many more design possibilities and make material selection more difficult.

Strain hardening is applicable to almost all metals; the only metals that do not strain harden are those that recrystallize at room temperature. Substantial increases in the yield and tensile strength can be obtained through strain hardening. One drawback to strain hardening is that the property increase is not uniform in the depth direction or in the planar direction along the surface. The differences in the planar direction can be somewhat controlled by proper annealing and rolling cycles.

There are two different types of heat treatment, allotropic transformation suppression and precipitation hardening, but they follow the same general procedure. The heat treatment procedure has three basic steps: heating to form a homogeneous solid solution; quenching to form a supersaturated solid solution; and reheating either to temper the material (martensite products) or to age the material (precipitation-hardenable materials).

4.7   EVALUATIVE QUESTIONS

1. If the tensile strength constant for a steel alloy was 40,000 psi and the solid solution concentration constant for chromium was 60,000 psi, what would be the increase in strength if the chromium level increased from 0.01 to 0.02 percent? from 0.01 to 0.04 percent? from 0.01 to 0.1 percent?

2. If the grain boundary strength constant is 400 psi/in. and the strength of a grain is 30,000 psi, what is the change in strength if the grain size goes from 0.001 in. to 0.003 in.?

3. An annealed aluminum wire is drawn through a die to reduce the diameter from 10 mm to 8 mm. What is the amount of cold work in the material after drawing?

4. A sheet of annealed steel is cold rolled from 10 mm thick to 8 mm thick. What is the amount of cold work? The material is further reduced to 6 mm. What is the additional amount of cold work, and what is the total cold work? Are the amounts of cold work additive?

5. Estimate the recrystallization temperature ranges for aluminum and steel if the melting temperatures are 660° and 1535° C. Use the general relationship between recrystallization temperature and melting temperature.

6. A sheet metal tensile specimen has the following original dimensions: 25 mm long, 10 mm wide, and 2 mm thick. After the load is applied, the final dimensions are 37 mm long and 8 mm wide. Assuming constant volume, calculate the strain ratio.

7. If the strain ratios for the directions 0°, 45°, and 90° are 0.9, 1.2, and 1.1, respectively, what are the values of the mean anisotropic coefficient and the planar anisotropic coefficient?

8. What are the two types of heat treatment? What are the restrictions for each of the methods?

9. What are the three steps for heat treatment? What is the difference in purpose of the last step for the two different types of heat treatment?

10. What are the differences between pearlite, bainite, and martensite?

REFERENCES

1. Barrett, C. R., Nix, W. D., and Tetelman, A.S. The Principles of Engineering Materials, Prentice-Hall, Englewood Cliffs, NJ, 1973, pp. 251–292.
2. ASM Handbook, Vol. 2. Properties and Selection: Nonferrous Alloys and Special Purpose Materials, ASM International, Metals Park, OH, 1990, pp. 49, 50, 265.
3. Tool and Manufacturing Engineers Handbook, Vol. 2:—Forming, 4th ed., 1984, Society of Manufacturing Engineers, Dearborn, MI, pp. 1–8, 1–10.
4. Keeler, S. P., "Understanding Sheet Metal Formability–Properties Related to Forming," Part 3, Machinery, April 1968.
5. Kalpakjian, S. Manufacturing Processes for Engineering Materials, 2nd ed., 1991, Addison-Wesley, Reading, MA, pp. 452–456.

BIBLIOGRAPHY

Faupel, Joseph H., and Fisher, Franklin E. Engineering Design, Wiley, 1981.
Juvinall, Robert C., and Marshek, Kurt M. Fundamentals of Machine Component Design, 2nd ed., Wiley, 1991.
Lewis, Gladius. Selection of Engineering Materials, Prentice Hall, Englewood Cliffs, NJ, 1990.

INTERNET SOURCES

Process and theory of heat treatment: *http://vims.ncsu.edu/Contents/Ch.12/Chapter12TOC.html*
Heat treating company facilities: *http://www.rafael.co.il/mtc-htf.html*
Superplastic materials: *http:www.mm.mtu.edu/~drjohn./superplasticity.html*

5
Material Codes and Coding Systems

5.1 INTRODUCTION

There are, on the conservative side, over 1 million different engineering materials for manufactured products. This would include all the various metals (crystalline and the new metal glasses), ceramics (including porous ceramics), polymers, glasses, elastomers, foams, woods, and composites (1). The composites would include ceramic matrix composites, polymer matrix composite, and metal matrix composites that are being developed for structural materials (2). In addition, numerous electronic materials and optical materials are being developed for the computer, communication, and information technologies; these are generally not considered as structural materials.

In addition to the vast number of materials, there are a wide variety of organizations creating codes for materials; thus, the same material may be classified differently and be considered as different materials. The focus of this chapter is engineering materials—in particular, metals. This is because materials are classified according to material properties as well as composition, and the metal systems are better defined than are the other systems. For example, strength is measured differently for different materials (1): metals and polymers use yield strength; ceramics and glasses use compressive strength; composites use tensile failure; and elastomers use tensile tear strength.

Materials comprise the major cost of most products; and for manufacturers to remain competitive, new materials are being developed to reduce product costs. As new materials are developed and their consumption increases, the material prices are reduced as a result of the economies of scale and process innovation.

5.2 GENERAL METAL CODING SYSTEM

The coding of materials, and in particular of metals, has become more important because there are more and more new alloys being developed and

Table 5.1   UNS Letter Prefix for Various Alloys

Letter prefix	Main metal or alloy group
A	Aluminum alloys
C	Copper alloys
E	Rare earth materials (cesium, cerium, etc.)
F	Cast irons
G	AISI and SAE carbon steels
H	Hardenable steels (AISI & SAE H-steels)
J	Cast steels (except tool steels)
K	Miscellaneous steels and ferrous alloys
L	Low-melting metals (bismuth, cadmium, tin, etc.)
M	Miscellaneous nonferrous (antimony, arsenic, etc.)
N	Nickel and nickel alloys
P	Precious metals
R	Reactive and refractory metals
S	Heat- and corrosion-resistant stainless steels
T	Tool steels
W	Welding filler metals
Z	Zinc and zinc alloys

the differences between materials are more significant in the highly competitive world marketplace. The Unified Numbering System (UNS) (3) was developed by the Society of Automotive Engineers (SAE) and the American Society for Testing and Materials (ASTM) to be comprehensive for all metal alloys and yet retain much of the existing coding methodologies. However, the Unified Numbering System has not been well received, especially in the steel and aluminum industries, and the former codes are still frequently used.

The Unified Numbering System is a code consisting of a letter prefix and a five-digit number. The letter prefix indicates the main metal or alloy group; see Table 5.1.

5.3   MATERIAL CODES AND CLASSIFICATIONS FOR FERROUS MATERIALS

The most common classification system for steels is the AISI-SAE classification, that is, the scheme developed by the American Iron and Steel Institute and the Society of Automotive Engineers. The AISI represented the suppliers of steel, and the SAE represented one of the largest steel consumer groups, the automotive industry, and it was natural for these two groups to develop a common classification system. The code consists of four digits, the first representing the main alloy or group and the last representing the carbon content, in one-hundredths of a percent. For instance, if the last two digits were 52, the carbon content would be 0.52 weight percent. The commonly used steel groups are presented in Table 5.2.

The UNS code is related to the AISI-SAE code for most steels. For example, the UNS code for AISI 1040 steel would be G10400; the differences are that another zero is added to the end, and G represents the AISI-SAE steels. If a new steel is added or a special alloy is used, the last digit would not be zero. For example, AISI-SAE 50B46 means that boron is added; the corresponding UNS code is G50461. The problem in revising a code is that people don't like to change; although the United States has officially been on the metric system for over 100 years, industry still prefers psi instead of MPa and Fahrenheit instead of Centigrade. Most companies use the AISI-SAE code rather than the UNS code.

There are several other coding systems for steel alloys. There is the letter coding for tool steels and the structural steel codes, developed by ASTM, which tend to have the letter *A* and three digits. These codes do not match the UNS codes. For example, A242 steel has the UNS designation of K11510. There are also general classifications of steels, such as high- or low-carbon steel and high-alloy or low-alloy steels. A general description of these limits is presented in Table 5.3; the values listed there are general guidelines. For example, in sheet deformation, low carbon may be 0.15 instead of 0.30 carbon.

Stainless steels, which have chromium levels greater than 12 percent, are high-alloy steels. Stainless steels are generally classified as austenitic, martensitic, and ferritic. The austenitic steels are not ferromagnetic and gen-

Table 5.2 AISI-SAE Main Alloy Groups for Steels

Code[a]	Main alloy or alloys
1XXX	Carbon (plain carbon)
2XXX	Nickel
3XXX	Nickel-chromium
4XXX	Molybdenum with chromium and nickel
5XXX	Chromium
6XXX	Chromium-vanadium
8XXX	Nickel-chromium-molybdenum

[a]XXX represents the last 3 digits of the code, which identify the specific alloy; the first digit indicates the main alloy(s).

Table 5.3 General Classifications of Steels

Carbon steels	Carbon content	Alloy steels	Total alloy content (%)
Low-carbon steel	<0.30	Low-alloy steels	<10 %
Medium carbon steel	0.30–0.60	High-alloy steels	>10 %
High carbon steel	>0.60		

erally have at least 24 percent alloy content. The martensitic and ferritic stainless steels are both ferromagnetic, and the martensitic steels generally require higher carbon levels to form the martensitic structure.

Stainless steels were traditionally classified as martensitic, ferritic, and austenitic. The austenitic stainless steels, which were coded as 2XX and 3XX (200 and 300 steels), have the austenitic structure at room temperature. These steels were very good at both low and high temperatures, but could be strengthened only by work hardening. The 304 stainless steel (one of the 18 percent Cr and 8 percent Ni steels) is classified as S30400 in the UNS code. The 4XX series was used for both martensitic and ferritic steels, and the differences were hard to distinguish by composition. The ferritic stainless steels have lower carbon levels (usually lower than 0.12 C) and higher chromium levels. In addition to the three traditional types, precipitation hardening stainless steels and duplex stainless steels have been developed.

Cast irons have much higher carbon levels than steels; they occur in the eutectic portion of the iron–iron carbide phase diagram, whereas steels are in the eutectoid regions of the iron–iron carbide phase diagram. Cast irons generally have carbon levels in excess of 2 percent; they can exceed 4 percent for some ingot mold irons, but the carbon level usually is between 3 and 4 percent. In addition to carbon, there is usually at least 1 percent silicon in the alloy, so cast irons are more of a ternary iron-carbon-silicon system rather than the binary iron-carbon system. Cast irons are called "cast" because they are used predominantly for metal castings; with their high carbon levels and brittle structures, they cannot easily be formed by the various deformation processes.

The properties of cast iron are greatly affected by the structure the carbon takes. In white iron, all of the carbon is in the form of carbides; and thus the material is very hard and brittle. In the other cast iron structures, some of the carbon appears as graphite or temper carbon. The matrix of the cast iron can be either ferrite, pearlite, bainite, or martensite, so the properties of cast irons can vary considerably as a result of the matrix. In addition, the carbon can take the form of thin flakes (gray iron), thick flakes

Table 5.4 Graphite Forms and Matrix Structures for Cast Iron Types

Cast iron type	Graphite form	Common matrix structures
White iron	No free graphite	Martensite, carbides
Gray iron	Graphite flakes	Ferrite, pearlite
Compacted iron	Thick (vermicular) flakes	Ferrite, pearlite
Malleable iron	Rosettes	Ferrite, pearlite
Nodular (ductile) iron	Spheroids	Pearlite, ferrite, bainite, martensite

(compacted or vermicular graphite in compacted iron), spheroids (nodules in ductile iron), or rosettes (temper carbon in malleable iron). The flakes make the structure more brittle, but they also improve the damping capacity and thermal conductivity of the material. The spheroids and rosettes greatly improve the ductility of the material and can be used with the harder matrix materials, such as martensite and bainite. Some of the most common matrix and graphite forms are presented in Table 5.4. The advantages of cast irons over steels are that they melt at lower temperatures and are easier to cast, but the advantage of steels is that they can easily be processed into sheet material and formed into a wide variety of products by sheet metal deformation processes.

Gray cast irons were generally classified according to mechanical properties under the ASTM system; for example, an ASTM class-20 gray iron meant that the alloy had a minimum tensile strength of 20 ksi (typical value of 22 ksi); correspondingly, a class-50 gray iron would have a minimum tensile strength of 50 ksi. Gray cast irons are easily machined because of the graphite flake structure throughout the material.

The ductile (nodular) irons were also classified according to mechanical properties under the ASTM system; for example, an ASTM A-395 ductile iron was classified as a 60-40-18 ductile iron, which meant that it had 60-ksi tensile strength, 40-ksi yield strength, and 18 percent elongation. The new UNS code is F32800, which is 328 MPa, equivalent to 40 ksi, and does not indicate the tensile strength or the elongation. Thus most design engineers tend to follow the old system, since it gives more design information.

5.4 CODING FOR ALUMINUM ALLOYS

The Aluminum Association (AA) has developed coding systems for aluminum materials, one code for fabrication and a second code for casting alloys. The most frequently used code is the fabrication code, summarized in Table

5.5. It is a four-digit code, with the first digit indicating the major alloy other than aluminum, if there is any. Some of the aluminum alloys are heat treatable with respect to forming precipitation hardening alloys, and this is also indicated in Table 5.5.

Thus the aluminum alloy code of 3003 H-14 means that the main alloy is manganese and that the alloy has been strain hardened by rolling to the half-hard condition. On the other hand, a 7075-T651 alloy code means that this aluminum alloy has zinc as the main alloying element and that it has been precipitation hardened according to the specific treatment process coded 651. The UNS code for 3003 is A93003 and for 7075 is A97075; the temper conditions, such as T-651 and H-14, are not in the basic UNS coding system.

Table 5.5   Summary of Aluminum Fabrication Code

Code	Major alloy element(s)	Alloys Heat Treatable?
A. Alloy composition codes		
1XXX	Aluminum (pure aluminum)	No
2XXX	Copper	Yes
3XXX	Manganese	No
4XXX	Silicon	No
5XXX	Magnesium	No
6XXX	Magnesium and silicon	Yes
7XXX	Zinc	Yes
8XXX	Other	Some
9XXX	Unused	Not applicable

Code	Condition
B. Temper codes	
O	Annealed
H	Strain hardened
F	Temper as fabricated
T	Heat-treated (tempered)

Code	Description
C. Strain-hardened codes	
H-12	Quarter-hard
H-14	Half-hard
H18	Fully hardened (75% reduction)

Adapted from Ref. 4.

In the Aluminum Association casting code, the code is of the general form XXX.X; thus, 356.0 is a typical die casting alloy. For the UNS system, the code would be A03560. The casting codes are A0XXXX and the fabrication codes are A9XXXX for the UNS system.

5.5 MATERIAL CODES AND PROPERTIES

The titanium, magnesium, and copper alloys are presented in Table 5.6, along with the aluminum and ferrous alloys. The copper alloys follow the code developed by the Copper Development Association (CDA). The CDA code for yellow brass is 853 and the UNS code is C85300. Similarly, the magnesium codes for the UNS system also follow those developed by the Magnesium Association. The UNS code for the magnesium die-casting alloy AZ91A is M11910. Table 5.6 also gives material properties for various materials, including some polymers as well as the ferrous and nonferrous alloys. The polymers were not included in the UNS system.

5.6 SUMMARY

There are over 1 million different materials that can be considered for mechanical or structural designs. Classification systems are required to identify the different materials. The Unified Numbering System is an attempt to classify all metal systems, and it has included many of the highlights of existing classification systems. One of the problems in comparing widely different materials is that the properties (such as strength) are not measured in the same manner, so care must be used in making comparisons. Codes for the two most common metals, steels and aluminum, are presented in detail.

5.7 EVALUATIVE QUESTIONS

1. a. What is the main alloy(s) and the carbon content of the following steels?

    i. 1015
    ii. 4140
    iii. 8640

  b. What is the main alloy(s) of the following aluminum alloys?
    i. 1100
    ii. 2024
    iii. 7075

Table 5.6 Material and Mechanical Properties[a] for a Variety of Alloys

General material description	Material code Typical	UNS	Density (lb/in.³)	Strength[b] (kpsi) Yield	UTS	Modulus (Mpsi)	Elongation (%)	Composition, major elements
A. Ferrous Alloys								
Plain-carbon and low-alloy steel								
Low Carbon	1010	G10100	0.29	26	47	30	38	C (0.08–0.13)
Low carbon	1020	G10200	0.29	30	55	30	35	C (0.17–0.23)
Medium carbon	1040	G10400	0.29	42	76	30	28	C (0.36–0.44)
Medium carbon (Q&T)	1040	G10400	0.29	110	85	30	18	C (0.36–0.44)
High carbon	1080	G10800	0.28	50	90	30	25	C (0.74–0.88)
High carbon	1095	G10950	0.28	55	95	30	20	C (0.90–1.04)
Low alloy	4140	G41400	0.29	60	95	30	25	C (0.4), Cr (0.95), Mo (0.2)
Low alloy	4340	G43400	0.29	65	105	30	22	C (0.4), Cr (0.8), Ni (1.8), Mo (0.25)
Low alloy	1340	G13400	0.29	60	100	30	25	C (0.3), Mn (1.75)
High-strength, low-alloy								
Low alloy	A242	K11510	0.29	45	65	30	18	C (0.15), Mn (1.0), Cu (0.2)
Low alloy	A618	K12609	0.29	50	70	30	18	C (0.22), Mn (1.1), Cu (0.2), V (0.02)
Ultrahigh-strength steels								
Low alloy (Q&T)	8640	G86400	0.29	180	200	30	10	C (0.4), Mn (0.9), Cr (0.5), Ni (0.6), Mo (0.2)
Low alloy	D6		0.29	240	260	30	8	C (0.4), Si (1.6), Cr (1.1), Ni (0.6), Mo (1.), V (0.2)
High alloy	18Ni (300)			290	300		6	Ni (18), Mo (4.2), Co (12.5), Al (0.1), Ti (1.6)

	Alloy	UNS No.						Composition, %
High-alloy steel								
Stainless—austenitic	202	S20200	0.28	55	100	29	55	C (0.15), Cr (18), Ni (5), Mn (8.5)
Stainless—austenitic	302	S30200	0.29	35	85	28	60	C (0.15), Cr (19), Ni (9), Mn (2)
Stainless—austenitic	304	S30400	0.29	30	75	28	40	C (0.08), Cr (19), Ni (9), Mn (2)
Stainless—austenitic	308	S30800	0.29	30	85	28	55	C (0.08), Cr (20), Ni (11), Mn (2)
Stainless—martensitic	410	S41000	0.28	30	65	29	22	C (0.15), Cr (12.5), Ni (0.75), Mn (1), Si (1)
Cast irons								
Gray iron Class 30			0.26		30	14	<1	
Gray iron Class 50			0.26		50	20	<1	
Ductile iron 60-40-18		F32800	0.26	40	60	24	18	
Ductile iron 80-55-06		F33800	0.26	55	80	24	6	
B. Nonferrous alloys								
Aluminum alloys								
Annealed	1100	A91100	0.098	5	13	10	35	Al (99.0)
Annealed	2024	A92024	0.10	11	27	10	20	Cu (4.4), Mg (1.5), Mn (0.6)
Heat treated	2024-T851	A92024	0.10	65	70	10	6	
Annealed	3003	A93003	0.098	6	16	10	30	Mn (1.3), Fe (0.7), Si (0.6), Cu (0.1)
Strain hardened	3003-H18	A93003	0.098	27	29	10	4	
Annealed	7075	A97075	0.10	15	33	10	17	Zn (5.5), Mg (2.5), Cu (1.6), Ti (0.2)
Heat treated	7075-T6	A97075	0.10	73	83	10	11	
Titanium alloys								
Ti-6Al-4V		R56400	0.16	120	130	16	10	Al (6), V (4)
Ti-6Al-2Sn-4Zr-6Mo		R56420	0.16	120	130	16	10	Al (6), Sn (2), Mo (6), Zr (2)

Table 5.6 Continued

General material description	Material code		Density (lb/in.³)	Strength[b] (kpsi)		Modulus (Mpsi)	Elongation (%)	Composition, major elements
	Typical	UNS		Yield	UTS			
Ti-13V-11Cr-3Al		R58010	0.17	160	170	16		Al (3), V (13), Mo (11)
Magnesium alloys								
Die-casting alloy	AZ91A	M11910	0.065	22	33	6.5	3	Al (9), Zn (0.7)
Extruded Mg	AZ80A-T5	M11800	0.065	40	55	6.5	7	Al (9), Zn (0.5)
Copper alloys								
High conductivity	805	C80500	0.32	9	25	17	40	Cu + Ag (99.75)
Commercial bronze	834	C83400	0.32	10	35	15	30	Zn (10)
Yellow brass	853	C85300	0.31	11	35	15	40	Zn (30)
Die-casting yellow brass	858	C85800	0.30	30	55	15	15	Zn (40), Sn (1), Pb (1)
High-strength yellow brass	863	C86300	0.28	83	119	14	18	Zn (25), Fe (3), Al (6), Mn (3)
Aluminum bronze	954	C95400	0.27	35	85	16	18	Al (11), Fe (4)
Copper nickel	964	C96400	0.32	37	68	21	28	Ni (30), Fe (0.9)
C. Polymers								
Glass fiber reinforced polymer (GFRP)			0.064	—	55	4	3	Polyester or epoxy, glass (55)
Acrylics (PMMA)			0.042	—	8	0.4	4	Acrylic
Acrylobutadienestyrene (ABS)			0.038	—	6	0.3	7	
Nylon			0.039	—	10	0.3	150	

[a]Mechanical properties are approximate values and values should be obtained from manufacturers. The values presented are generally conservative. Sources used to estimate properties were ASM Metals Handbook (1961); ASM Metals Handbook (Vol. 1, 1990); ASM Handbook (Vol. 2, 1990); Smithells Metals Reference Book (1962); Metal Progress Databook (1960); Metals Handbook—Desk Edition (1985); Aluminum Standards and Data (1979); Copper Standards Handbook (1970); The Fundamentals of Iron and Steel Castings (1957); Iron Castings Handbook (1981); Cambridge Materials Selector (CMS 2.0, 1994).

[b]Strength values are for annealed conditon unless otherwise noted.

2. Find the composition and composition range limits for the following materials, and list your source of information.

Steels	Aluminum alloys	Stainless steels
a. 1040	a. 2024	a. 202
b. 4340	b. 3003	b. 308
c. 4140	c. 6063	c. 316
	d. 7075	

3. Find the compositions and/or mechanical property ranges for the following cast iron materials, and list your source of information.
   a. ASTM class-20 gray cast iron
   b. ASTM class-50 gray cast iron
   c. 60-40-18 ductile iron
   d. 120-90-02 ductile iron
   e. 32510 malleable iron (UNS)
   f. 50005 malleable iron (UNS)

REFERENCES

1. Ashby, M. F. Materials Selection in Mechanical Design, Pergamon Press, 1992, p. 311, and accompanying Materials and Process Selection Charts, p. 57.
2. Office of Technology Assessment, Advanced Materials by Design, U.S. Government Printing Office, Washington, DC, June 1988, p. 335.
3. Society of Automotive Engineers, Metals & Alloys in the Unified Numbering System, 5th ed., 1989.
4. Aluminum Standards and Data-1979, The Aluminum Association, March 1979, Washington DC, p. 7–14.

BIBLIOGRAPHY

Charles, J. A., and Crane, F. A. A. Selection and Use of Engineering Materials, 2nd ed., 1989. Butterworths, Stoneham, MA.
Lewis, Gladius. Selection of Engineering Materials, Prentice Hall, Englewood Cliffs, NJ, 1990.

INTERNET SOURCES

Stainless steels: *http://www.arcus.nl/cat/p61_62htm*

6
Design, Material, and Cost Relationships

6.1 INTRODUCTION

The integration among product functional design, materials selection, manufacturing processes selection, and product cost has become necessary as the marketplace has become more internationalized and more competitive. This has led to the integration of economics, material properties, manufacturing processes, and engineering design relationships. Mathematical expressions have been developed that can integrate design condition, material property, and material cost data to help in the selection of materials.

The expressions developed will be for simple tension loading and for restrictions of load and elongation. Expressions can be developed for more complex loading, such as bending or torsion, but the purpose here is to illustrate the methodology and not to become overwhelmed by the calculations. Design problems at the end of the chapter permit students to develop the relationships for some bending situations.

6.2 BASIC EXPRESSION DEVELOPMENT

The procedure starts with the general expression to determine the material cost and then goes on to modify this expression to include parameters that reflect the material properties and design requirements (1). This expression is:

$$C(u) = C(w) \times W \tag{6.1}$$

where

$C(u)$ = total unit cost, \$
$C(w)$ = cost per unit weight, \$/weight (\$/kg, \$/lb)
$\quad W$ = weight (kg, lb)

The material weight can be expressed in terms of the material density and volume of material, so Eq. (6.1) can be rewritten as:

$$C(u) = C(w) \times d \times V \tag{6.2}$$

where

d = density, weight/unit volume (kg/m^3, lb/in.^{3})
V = volume (m^3, in.^{3})

This brings the material property of density into the evaluation of the cost. The volume can be expressed in terms of the cross-sectional area and length; the cross-sectional area is frequently involved in the basic design relationship. In general, the design length would be a specified design parameter, whereas the cross-sectional area would be a design variable. The new expression is:

$$C(u) = C(w) \times d \times L \times A \tag{6.3}$$

where

L = design length (m, in.)
A = design cross-sectional area (m^2, in.^{2})

The cross-sectional area is usually determined by the design requirements. Two of the most common design requirements are that the product support a design load (strength requirement) and that the product be restricted in the amount of deflection (stiffness requirement).

6.3 MINIMUM-COST DESIGN FOR STRENGTH REQUIREMENTS FOR SIMPLE TENSION LOADING

The design relationships for the cross-sectional area must be developed to apply Eq. (6.3). The first step is to obtain the basic design relationships and then to convert them into the form required. The basic design relationships for strength, for stiffness, and for different types of loading and cross-sectional shapes are presented in Table 6.1.

For the case of simple tension in a solid bar, the design relationship for strength, from Table 6.1, is:

$$S = P/A \tag{6.4}$$

where

S = material strength (MPa, kpsi)
P = load (kN, lb)
A = cross-sectional area (m^2, in.^{2})

Table 6.1 Basic Design Relationships for Various Loading Types and
Cross-Sectional Shapes

Structural cross-sectional shape	Type of load	Design relationship for:	
		Strength	Stiffness
Solid bar, rod or cylinder	Tension, compression	$S = P/A$	$\delta = PL/AE$
Solid cylinder	Bending (center load)	$S = mc/I$	$\delta = PL^3/48EI$
Solid rectangular plate	Bending (center load)	$S = mc/I$	$\delta = PL^3/48EI$

The relationship can be solved for the area by rearranging the expression and obtaining:

$$A = P/S \tag{6.5}$$

If Eq. (6.5) is used in Eq. (6.3), the expression for the unit cost can be written as:

$$C(u) = (P \times L) \times [C(w) \times d/S] \tag{6.6}$$

where the first part of the expression represents the specific design requirements, (P, L) and the second part of the expression represents the material costs and properties $[C(w), d,$ and $S]$. Since the first part of the expression is fixed for a specific design, the second part of the expression can be used to determine the best or optimal material for the minimum cost. If the second part of the expression is minimized, then the total cost of the product would be minimized, since the first expression is a constant. This second expression is often called the *cost performance ratio*; the lower the ratio, the lower the unit cost will be for the conditions evaluated.

The material property data required for many of the various performance ratios is presented in Table 6.2. The cost performance ratio can be calculated for the material and used to select the best material for the particular type of loading. The cost performance ratio for simple tension or compression for a solid rod, bar, or cylinder is:

$$PR(1) = C(w) \times d/S \tag{6.7}$$

where

$PR(1)$ = strength cost performance ratio (for simple tension or compression in a solid rod, bar, or cylinder)
$C(w)$ = cost per unit weight
S = yield strength
d = density

Table 6.2 Material Property and Cost Data

Material description	Yield strength		Density		Elastic modulus		Material cost	
	MPa	(kpsi)	kg/m³	lb/in.³	GPa	(Mpsi)	$/kg	($/lb)
Magnesium								
AS91C-T6	145	(21)	1831	(0.066)	44.8	(6.5)	4.40	(2.00)
AK60A-T5	283	(41)	1831	(0.066)	44.8	(6.5)	4.84	(2.20)
Aluminum								
2024-T861	455	(66)	2774	(0.100)	72.4	(10.5)	2.20	(1.00)
3003-H18	200	(29)	2746	(0.099)	70.3	(10.2)	2.00	(0.91)
7175-T66	524	(76)	2802	(0.101)	71.7	(10.4)	2.75	(1.25)
Titanium								
Ti-6Al-6V-2Sn	965	(140)	4531	(0.165)	110	(16.0)	13.20	(6.00)
Ti-13V-11Cr-3Al	1103	(160)	4854	(0.175)	110	(16.0)	13.75	(6.25)
Steels								
1015	345	(50)	7905	(0.285)	207	(30.0)	0.60	(0.27)
4140 Q&T	1738	(252)	7905	(0.285)	207	(30.0)	2.75	(1.25)
304 Stainless	586	(85)	7961	(0.287)	193	(28.0)	4.40	(2.00)
Cast Iron								
Gray Cast Iron	138	(20)	6935	(0.250)	103	(15.0)	0.40	(0.18)
Ductile Iron	345	(50)	6935	(0.250)	207	(30.0)	0.50	(0.23)

Material property data is from Refs. 2a–2c, and is reprinted with permission of ASM International. The cost data is estimated.

The performance ratios for the various materials in Table 6.2 are presented in Table 6.3. The performance ratios are dependent upon the cost per unit weight, $C(w)$, and must be updated for changes in material costs. The other material properties, such as density, strength, and elastic modulus, will generally not change.

A second performance ratio, the stiffness cost performance ratio, for a member in simple tension or compression is determined by:

$$PR(2) = C(w) \times d/E \qquad (6.8)$$

where

PR(2) = stiffness cost performance ratio (for simple tension or compression in a solid rod, bar, or cylinder)
$C(w)$ = cost per unit weight
E = elastic modulus
d = density

The stiffness cost performance ratios are also included in Table 6.3 in both metric and USA units. Materials with the smallest ratios have the best performance on a cost basis. For the materials in Table 6.3, ductile iron has the best strength cost performance ratio and the best stiffness performance cost ratio. However, aluminum alloys 2024-T861 and 7075-T66 and steels such as 1015 and 4140 Q&T are competitive with respect to the cost strength

Table 6.3 Cost Performance Ratios for Materials

Material	Strength performance cost ratio—PR(1)		Stiffness performance cost ratio—PR(2)	
	Metric units[a]	USA units[b]	Metric units[a]	USA units[b]
Aluminum				
2024-T861	13.4	1.51	84.3	9.52
3003-H18	29.5	3.11	78.1	8.83
7075-T66	14.7	1.66	107	12.1
Steel				
1015	13.7	1.54	22.9	2.57
4140 Q&T	12.5	1.41	105	11.9
304 stainless	59.8	6.75	181	20.5
Cast iron				
Gray cast iron	20.1	2.25	26.9	3.00
Ductile iron	10.1	1.15	16.8	1.92

[a]Values used directly from Table 6.2 with no adjustment.
[a]Values used directly from Table 6.2 and multiplied by 1000.

performance ratio; this means that the cost data must be updated and accurate. When stiffness cost performance is considered, gray cast iron and 1015 steel are competitive with the ductile iron.

Other performance ratios can be utilized for cases when cost is not the primary criterion for that component. This situation can arise in aircraft structures, where weight is critical and any added weight will increase the cost of other components more than the cost of the component under investigation. Thus two other performance ratios occasionally used are strength per unit weight and stiffness per unit weight. High values of these performance ratios, rather than low values, indicate good performance. The strength performance ratio is determined by:

$$PR(3) = S/d \qquad\qquad (6.9)$$

where

> PR(3) = strength performance ratio (simple tension or compression
> loading for a solid rod, bar, or cylinder)
> S = yield strength
> d = density

Similarly, a stiffness performance ratio can be developed for materials where component cost does not correctly represent the overall system cost:

$$PR(4) = E/d \qquad\qquad (6.10)$$

where

> PR(4) = stiffness performance ratio (simple tension or compression
> loading for a solid rod, bar, or cylinder)
> E = elastic modulus
> d = density

Table 6.4 presents the strength performance ratios and the stiffness performance ratios for the materials in Table 6.2. The values for the best materials have the highest ratios. Note that on a strength performance basis, the 4140 Q&T is the best material, with the 7075-T66 and 2024-T861 aluminum alloys only slightly lower. The ductile cast iron, which was the best on the strength, cost performance ratio data, has a lower value on the strength performance ratio than do all of the aluminum alloys considered, as well as the 4140 Q&T steel and 304 stainless steel.

The stiffness performance ratio values are fairly constant for the various materials, and the variation is much lower than that for the strength performance ratio. Thus, although the ductile iron is still the best, all of the materials are nearly equivalent in stiffness performance.

Table 6.4 Strength Performance Ratio and Stiffness Performance Ratio for Various Materials

Material	Strength performance ratio—PR(3)		Stiffness performance ratio—PR(4)	
	Metric units	USA units	Metric units	USA units
Aluminum				
2024-H861	0.164	660	0.026	105
3003-H18	0.073	293	0.026	103
7075-T66	0.187	752	0.026	103
Steel				
1015	0.044	175	0.026	105
4140 Q&T	0.220	884	0.026	105
304 stainless	0.074	296	0.024	98
Cast iron				
Gray cast iron	0.020	80	0.022	60
Ductile iron	0.050	200	0.030	120

All values calculated from data in Table 6.2 with no adjustment.

6.4 DESIGN PERFORMANCE CALCULATIONS FOR OTHER TYPES OF LOADING

The basic expression for the evaluation of performance is Eq. (6.3), which is repeated here:

$$C(u) = C(w) \times d \times L \times A \tag{6.3}$$

The key to the use of this equation for other types of loading is the calculation of the design cross-sectional area. In the case of simple tension, the expression was easily evaluated as P/S, as indicated by Eq. (6.5). However, for more complex loading, such as bending in a flat plate or cylinder, the expression for the area is more complex. The design relationship for strength for a solid rectangular plate, as indicated in Table 6.1, is:

$$S = mc/I \tag{6.11}$$

where

m = bending moment
c = centroid
I = moment of inertia

If the rectangular plate is considered to have length L, width w, and thickness t, and the L and w values are fixed and only t is permitted to vary, the values of m, c, and I can be expressed as:

$$m = PL/4$$
$$c = t/2$$
$$I = wt^3/12$$

where

P = load applied at midpoint of plate
L = length between supports
w = width of plate
t = thickness of plate

Thus S can be expressed as:

$$S = \frac{PL/4 \times t/2}{wt^3/12}$$
$$= PL \times 3/2wt^2 \tag{6.12}$$

but if w is fixed and $A = wt$, the expression can be rewritten as:

$$S = 3PLw/2A^2 \tag{6.13}$$

and A can be evaluated as:

$$A = (3PLw/2S)^{1/2} \tag{6.14}$$

When this value of A is used in Eq. (6.3), the expression becomes:

$$C(u) = C(w) \times d \times L \times (3PLw/2S)^{1/2}$$
$$= C(w) \times d/S^{1/2} \times (3PL^3w/2)^{1/2} \tag{6.15}$$

where the first part of the expression is the new strength cost performance ratio; that is:

$$PR(1a) = C(w) \times d/S^{1/2} \tag{6.16}$$

where

$PR(1a)$ = strength cost performance ratio (for simple bending in a solid bar or plate)
$C(w)$ = cost per unit weight
d = density
S = yield strength

It is important to note that the exponent of the strength term is only one-half of that in Eq. (6.7). This is an important difference, and the problem in evaluating the performance ratios for materials are that they are dependent upon the type of loading. Thus the ranking of materials for strength cost performance may be different for tension than for bending.

6.5 DESIGN CALCULATIONS WITH MULTIPLE CONSTRAINTS

When there are multiple constraints, such as both load and elongation, one must calculate the critical cross-sectional area rather than use cost performance ratios. The cost performance ratios are comparable for the same critical loading; but if load is critical for one material and elongation is critical for another material, the ratios *should not* be used.

An example problem will illustrate the difficulty with cost performance ratios and the use of critical cross-sectional areas.

Example Problem 6.1. Let two materials, 2024-T861 aluminum and 1015 steel, be considered for loading of a circular bar in simple tension that is 25 cm long (0.25 m) with a load of 200 kN and a maximum extension of 1 mm (0.001 m). The data of Table 6.2 will be utilized. If we use the material cost performance ratios in Table 6.3, then the steel is better than the aluminum for stiffness but the aluminum is slightly better than the steel for strength. Thus we need to evaluate the cross-sectional areas and then evaluate the total cost. The cross-sectional areas for the two materials for the two conditions of load and extension are presented in Table 6.5.

From the results in Table 6.5, the critical area of the aluminum (6.91×10^{-4} m^2) is determined by the extension whereas the critical area (5.80×10^{-4} m^2) of the steel is controlled by the load. The unit costs can be evaluated from Eq. (6.3) as:

$$C(u\text{-Al}) = 2.2 \text{ \$/kg} \times 2774 \text{ kg/m}^3 \times 0.25 \text{ m} \times 6.91 \times 10^{-4} \text{ m}^2$$
$$= \$1.05/\text{unit}$$
$$C(u\text{-Steel}) = 0.60 \text{ \$/kg} \times 7905 \text{ kg/m}^3 \times 0.25 \text{ m} \times 5.80 \times 10^{-4} \text{ m}^2$$
$$= \$0.69/\text{unit}$$

The total unit cost values for the constraints considered indicate that the steel is a much better material when both conditions (loading and extension) are considered for the cost values presented.

If the cost of aluminum were reduced to $1.20/kg, the cost performance ratios would change magnitude, but aluminum would still be better for strength and steel would still be better for extension. The areas would be the same, because they are independent of the cost, but the cost value for aluminum would be only $ 0.58 per unit and thus it would be preferred over the steel.

Table 6.5 Critical Cross-Sectional Areas for Example Problem 6.1

Material	Load area (m^2) $A = P/S$	Stiffness area (m^2) $A = PL/\delta E$
2024-T861 Al	4.40×10^{-4}	6.91×10^{-4}
1015 Steel	5.80×10^{-4}	2.42×10^{-4}

These calculations indicate that care must be exercised in making economic calculations when multiple constraints are involved with different types of loading. The straightforward approach of calculating the critical cross-sectional area is the easiest approach and will indicate the differences of what is critical for the different materials. This approach also indicates a method for determining what material costs are required for economical substitution of alternative materials.

6.6   ASHBY PLOTS FOR MATERIALS EVALUATION

A new and exciting approach for materials selection with design considerations is that by Ashby (3,4) at the University of Cambridge in England. He has developed a series of charts (currently 25) relating various material, mechanical, and process properties for a wide range of materials. Figure 6.1 indicates the strength of a material versus the relative cost per unit volume. The three guidelines are for different loading conditions:

a. Strong ties (simple tension) and rotating discs
b. Strong beams and shafts
c. Strong plates

More details on application of the charts is a topic for advanced study and research and beyond the scope of this book. Recently his research group has developed a computer program, CMS (Cambridge Materials Selector), that permits extensive evaluation of different materials. It also permits the comparison of the materials by different ratios, such as the strength cost performance ratio and the stiffness cost performance ratio. Other material properties, such as thermal, magnetic, and electrical properties, and other mechanical properties, such as hardness, fatigue, or stress intensity factors, can be used in the evaluation of materials. New material property ratios can easily be developed by the user for evaluation.

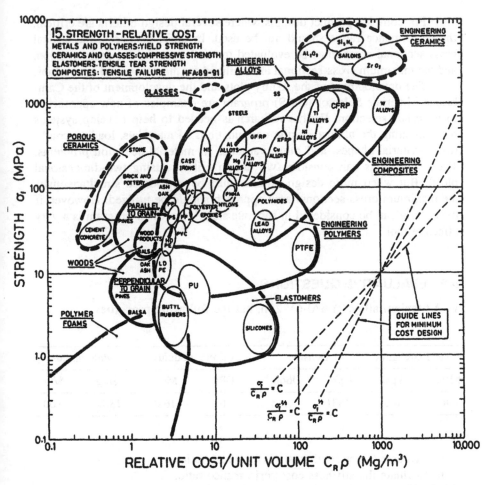

Figure 6.1 Ashby plot of strength versus relative cost per unit volume. (Used with permission, from Ref. 4.)

6.7 SUMMARY

Material properties are a critical parameter needed for the design of manufactured products. The type of loading and the design constraints control the critical cross-sectional areas, but what is critical is not the same for all materials when a variety of materials are being evaluated. When a single design criterion is dominant, the use of performance ratios can permit the rapid evaluation of a wide variety of materials.

For complex product designs, where several design criteria are critical, the performance ratios should not be used. In those instances, the critical cross-sectional areas must be evaluated for each material for each criterion, and the maximum cross-sectional area utilized for the design and cost values.

The development of the Ashby plots and the development of the Cambridge Materials Selector (CMS) program are examples of new approaches to materials selection. Further research is needed to help develop systems for evaluating the almost infinite combinations of materials, loading conditions, material shapes, and other variables in material selection problems. The Ashby plots can consider two design criteria (using two-dimensional plots). But when more design criteria must be considered simultaneously, the maximum cross-sectional area approach should be utilized. However, if the criteria can be considered in separate stages, the CMS program is a very effective tool.

6.8 EVALUATIVE QUESTIONS

1. A titanium alloy, Ti-6Al-6V-2Sn, has the following properties:

Yield strength		Density		Elastic modulus		Material cost	
MPa	kpsi	kg/m^3	lb/in.^{3}	GPa	Mpsi	$/kg	$/lb
965	140	4531	.165	110	16.0	13.20	6.00

    a. Evaluate the strength cost performance ratio.
    b. Evaluate the stiffness cost performance ratio.
    c. Evaluate the strength/weight performance ratio.
    d. Evaluate the stiffness/weight performance ratio.

2. Show that the stiffness cost performance ratio for bending of a bar or plate is $C(w) \times d/E^{1/3}$

3. Show whether the titanium alloy in Question 1 is better than the two materials for Example Problem 6.1. If the titanium alloy is not better, what would its maximum cost have to be to make it the best material?

4. Three materials are being considered for a design. The material design requirements specify that the part must support a load of 140,000 lb in simple tension and that the part have a maximum extension limit of 0.035 in. over the length of 10 in. The data for the materials being considered is:

Material	Yield strength YS (kpsi)	Elastic modulus E (Mpsi)	Density (lb/in.³)	Cost ($/lb)
A	30	10	0.066	6
B	50	15	0.100	5
C	90	20	0.275	3

a. What is the minimum material cost for each of the materials that meets the design requirements?
b. What is the weight of material used to meet the design requirements at minimum cost?
c. What is the volume of the material used to meet the design requirements at minimum cost?
d. How much must the unit cost of material A be reduced to make it the lowest-cost material meeting the design requirements?

6.9 RESEARCH QUESTIONS

1. Select one of the charts by Ashby and illustrate its application in the selection of a material for a specific design. Illustrate how the design guidelines are used.

2. Develop a computer program to take raw material data to generate a chart similar to those by Ashby.

3. Use the CMS (Cambridge Materials Selector) program to evaluate materials for one of the four design problems in Section 6.10, but use the CMS materials database.

6.10 MANUFACTURING DESIGN PROBLEMS

Design Problem 1. A rectangular beam is loaded at the center as indicated in the sketch in Fig. 6.2. The following design parameters are given:

$P = 15,000$ lb (design load)
$L = 150$ inches
$w = 4$ inches
$\delta = 2$ in. (max)

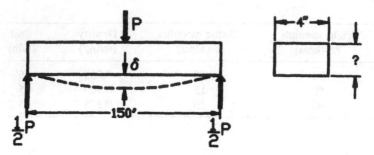

Figure 6.2 Rectangular beam with center load.

Assume that the weight of the structure is small with respect to the design load.

After considering all the materials in Table 6.2, determine the following.

a. The material that meets the design load and deflection limit at the lowest cost. Show all calculations.

b. The material that meets the design load and deflection limit with the lowest weight. Show all calculations.

c. The material that meets the design load and deflection limit with the lowest volume of material. Show all calculations.

d. If the design load is doubled, what is the effect on your calculations?

e. If the deflection limit is halved, what is the effect on your calculations?

The engineering report for this assignment should include, but not necessarily be limited to, the following:

a. A bound report

b. A one-page summary sheet describing the problem and the results

c. Explanation of any formulas used

d. Tables including the results of calculations made and explanation of the meaning of the calculations

e. A listing of not only the best material but also of the best three alternatives

f. References used: books, articles, etc.

g. Acknowledgement of human resources utilized: other students, professors, industry sources, library sources, etc.

h. Summary and conclusions about design and material selection

   Design Problem 2.   A circular cross-sectional beam is loaded at the center as indicated in the sketch in Fig. 6.3. The following design parameters are given:

   $P = 20,000$ lb (max)
   $L = 100$ in.
   $\delta = 3$ in. (max)

Assume that the weight of the structure is small with respect to the design load. After considering all of the materials in Table 6.2, determine the following.

a. The material that meets the design load and deflection limits at the lowest cost. Show all calculations, and include the diameter of the bar in the calculations.

b. The material that meets the design load and deflection limit with the lowest weight. Show all calculations.

c. The material that meets the design load and deflection limits with the smallest volume. Show all calculations.

d. If the design load is doubled, what is the effect upon your calculations?

e. If the deflection limit is halved, what is the effect upon your calculations?

   The engineering report for this assignment should include, but not be limited to, the following:

a. A bound report with appropriate cover

b. A one-page summary sheet describing the problem and the results obtained

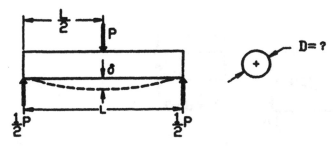

Figure 6.3   Circular beam with center load.

c. An explanation of the formulas used

d. Tables including the results of the calculations made and an explanation of the calculation results

e. A listing not only of the best material but also of the best three alternatives

f. References used: books, articles, etc.

g. Human resources used: other students, professors, graduate students, industry sources, library sources, etc.

h. A summary and conclusions about design and material selection

Design Problem 3. A square beam is loaded at the center as indicated in the sketch in Fig. 6.4. The following design parameters are given:

$$P = 15,000 \text{ lb (max)}$$
$$L = 150 \text{ in.}$$
$$\delta = 2 \text{ in. (max)}$$
$$\text{beam width} = \text{beam depth}$$

Assume that the weight of the structure is small with respect to the design load. After considering all of the materials in Table 6.2, determine the following.

a. The material that meets the design load and deflection limit at the lowest cost. Show all calculations, and include the dimensions of the bar.

b. The material that meets the design load and deflection limit at the lowest weight. Show all calculations.

c. The material that meets the design load and deflection limit at the lowest volume. Show all calculations.

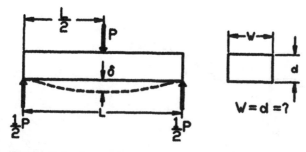

Figure 6.4   Square beam with center load.

d. If the design load is doubled, what is the effect upon your calculations?

e. If the deflection limit is halved, what is the effect upon your calculations?

The engineering report for this assignment should include, but is not limited to, the following:

a. A bound report with appropriate cover

b. A one-page summary sheet describing the problem and the results obtained

c. An explanation of the formulas used

d. Tables including the results of the calculations made and an explanation of the calculation results

e. A listing not only of the best material but also of the best three alternatives

f. References used: books, articles, etc.

g. Human resources used: other students, professors, graduate students, industry sources, library sources, etc.

h. A summary and conclusions about design and material selection

Design Problem 4. A triangular-shaped beam is loaded in simple tension as indicated in the sketch in Fig. 6.5. The following design parameters are given:

$P = 10,000$ lb (max)
$L = 20$ in.
$\delta = 0.1$ in. (max)
Triangle is equilateral

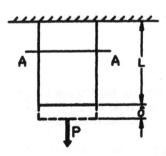

Figure 6.5 Triangular beam with center load

Assume that the weight of the structure is small with respect to the design load. After considering all of the materials in Table 6.2, determine the following.

a. The material that meets the design load and deflection limit at the lowest cost. Show all calculations, and include the dimensions of the bar.

b. The material that meets the design load and deflection limit at the lowest weight. Show all calculations.

c. The material that meets the design load and deflection limit at the lowest volume. Show all calculations.

d. If the design load is doubled, what is the effect upon your calculations?

e. If the deflection limit is halved, what is the effect upon your calculations?

   The engineering report for this assignment should include, but is not limited to, the following:

a. A bound report with appropriate cover

b. A one-page summary sheet describing the problem and the results obtained

c. An explanation of the formulas used

d. Tables including the results of the calculations made and an explanation of the calculation results

e. A listing not only of the best material but also of the best three alternatives

f. References used: books, articles, etc.

g. Human resources used: other students, professors, graduate students, industry sources, library sources, etc.

h. A summary and conclusions about design and material selection.

REFERENCES

1. Creese, R. C., Adithan, M., and Pabla, B. S. Estimating and Costing for the Metal Manufacturing Industries, Marcel Dekker, New York, 1992, pp. 86–95.
2a. ASM Metals Handbook, Volume 1: Properties and selection: Irons and Steels, 9th ed., 1978.
2b. ASM Metals Handbook, Volume 2: Properties and Selection: Nonferrous Alloys and Pure Metals, 9th ed., 1979.

2c. ASM Metals Handbook, Volume 3: Properties and Selection: Stainless Steels, Tool Materials and Special-Purpose Metals, 9th ed., 1980.
 3. Ashby, M. F. Materials Selection in Mechanical Design, Pergamon Press, New York, 1992, p. 311.
 4. Ashby, M. F. Materials Selection in Mechanical Design—Materials and Process Selection Charts, Pergamon Press, New York, 1992, p. 36.

BIBLIOGRAPHY

Charles, J. A., and F. A. A. Crane. Selection and Use of Engineering Materials, 2nd ed., Butterworths, 1989.
Dieter, George E. Engineering Design—A Materials and Processing Approach, 2nd ed., McGraw-Hill, Inc., 1991.
Faupel, Joseph H., and Fisher, Franklin E. Engineering Design, Wiley, New York, 1981.
Juvinall, Robert C., and Marshek, Kurt M. Fundamentals of Machine Component Design, 2nd ed., Wiley, New York, 1991.
Lewis, Gladius. Selection of Engineering Materials, Prentice Hall, Englewood Cliffs, NJ, 1990.

INTERNET SOURCES

Cambridge Materials Selector: *http://web.staffs.ac.uk/sands/engs/des/aids/materials/cms.htm*
Design Aids: *http://web.staffs.ac.uk/sands/engr/des/aids/aids.htm*

III
MANUFACTURING PROCESSES

7
Manufacturing Processes Overview

7.1 GENERAL PROCESS OVERVIEW

Manufacturing processing is the procedure by which materials are formed into the desired shapes. Figure 7.1 indicates the typical forms and shaping operations for the basic materials, such as metals, polymers, ceramics, and wood. Materials are formed into preliminary shapes, near net shapes, and then the final shaping and finishing operations are performed. Final treatments for material property improvement or surface improvement and assembly operations are the final stages in the manufacturing process.

Shaping processes can be classified into two main categories: primary shaping processes and secondary processing. The primary shaping processes form the overall shape of the product or of the components that will be joined to form the final product shape. The most common primary shaping processes are casting, bulk deformation processes such as forging and extrusion, sheet metal working, and machining. Other primary shaping processes include powder metallurgy and slush casting; these are sometimes included as casting or forging processes. One process, machining, is also frequently used as a secondary step for the other primary shaping processes. Thus, machining is used to improve the basic shape produced by casting or deformation processing as well as to produce the basic shape. Machining is the only process considered to be both a primary shaping process and a secondary shaping process.

The purpose of the secondary processing is to provide the final shape surfaces to meet some of the product requirements such as surface or dimensional tolerances. There frequently are several secondary shaping processes necessary to meet the product specifications. The emphasis of this chapter is to present an overview of the basic manufacturing processes. The following chapters will discuss the process selection basics and the basic manufacturing processes in more detail. For optimal product design, the

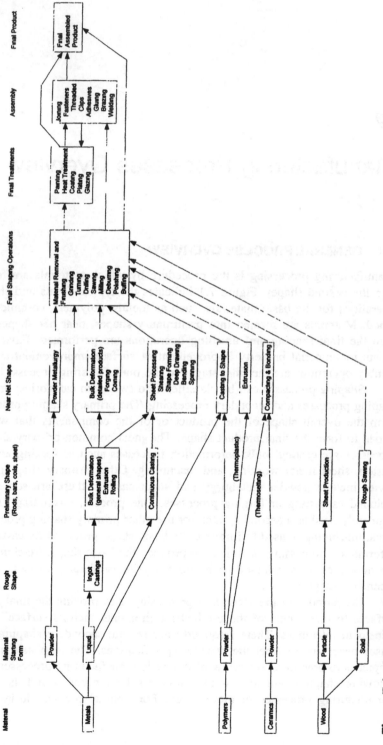

Figure 7.1 Manufacturing processing overview.

Table 7.1 Primary Component Processing Operations and Secondary Processing Operations

Primary operations		Secondary operations	
Primary shape formation	Primary shape completion	Assembly/ joining	Finishing
Casting (liquid metals and semisolid polymers)	Machining (component to final shape)	Fasteners	Tolerance requirements
Powder processing (metals/ ceramics in powder form)		Adhesives	Corrosion protection
Deformation processing (metals in bulk or sheet form)		Welding	Decorative painting & plating
Machining (chip/ chipless)			Packaging

optimal primary shaping and secondary processing operations must be selected together and not independently. Table 7.1 indicates the general focus of primary shaping processes and secondary finishing operations.

The selection of the primary shaping process is critical in controlling overall product costs. The shaping process selection is dependent upon the material and the product design. Both the shaping process and the material affect the product design, and this makes manufacturing so difficult to optimize. The primary manufacturing processes considered are casting, deformation processing (both bulk deformation and sheet metal deformation), machining (both chip generating processes and chipless processes), and powder processing. The discussion of the processes will emphasize metal materials, because most processes have been applied to metals, but similar analyses can be extended to other materials, such as ceramics, polymers, glasses, wood, and composites of these and other materials. Table 7.2 indicates the material state and how it varies as one proceeds from casting to deformation processes to machining. Some processes, such as slush casting and powder metallurgy, have some characteristics of both casting and forming.

Table 7.2 Primary Manufacturing Processes and Material State

	Manufacturing process						
Process:	Casting		Powder	Forming		Machining	
Subprocess:	Traditional casting	Slush casting	Powder metallurgy	Hot bulk	Sheet metal	Chip forming	Chipless
Material state:	Liquid	Liquid + solid	Powder	Hot solid	Thin sheet	Cold solid	Cold solid

7.2 CASTING OVERVIEW

Casting is one of the oldest and best-known manufacturing processes. It consists of pouring a liquid material, usually a metal, into a mold and letting it cool and solidify into the desired shape. The pouring and mold system is illustrated in Fig. 7.2. The making of gelatin molds is similar to the pouring,

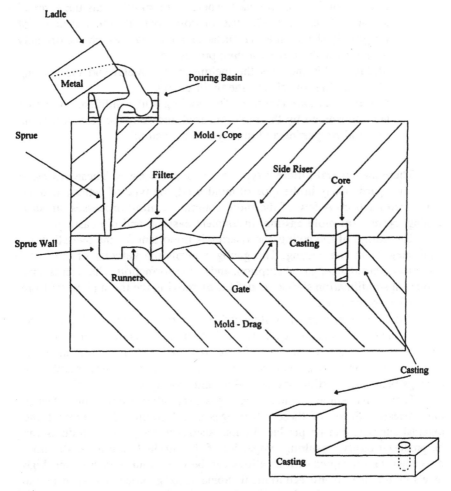

Figure 7.2 Pouring and mold system for a sand casting.

cooling, and solidifying operations of metal casting. There are four major areas of the casting process.

1. *Patternmaking*: This includes the modification of the desired part shape to include draft, shrinkage, and machining allowances as well as to provide the gating and risering design. The pattern has the shape of the casting, and it is used to make the cavity in the mold into which the metal is poured.

2. *Molding and coremaking*: This includes the production of the molds and cores, the insertion of the cores in the mold, and the closing of the mold. Cores typically are sand masses that form the internal surfaces of the part, whereas the mold forms the external surface of the part. The use of cores permits the formation of irregularly shaped internal surfaces in castings, which is extremely difficult for the other forming processes.

3. *Melting*: This includes the melting of the metal and the pouring of the molten metal into the mold cavity.

4. *Cooling and solidification*: The cooling and solidification, which are determined by the mold material and riser location, affect the material properties and quality of the casting.

There are several different types of casting processes. These processes may differ particularly in the type of mold used, the type of pattern, and the type of production. A few of the more common casting processes are sand casting, permanent mold casting, investment casting, and die casting.

The design of the gating and risering is important for the overall structural integrity of the casting. The gating controls the velocity of the metal and affects turbulence, gas absorption, and oxide formation. The riser is used to prevent solidification shrinkage defects by moving the thermal center from the casting to the riser.

The cost of castings is influenced by a large number of factors. The primary factors are casting weight, yield, melt loss, casting scrap rate, finishing scrap rate, and the amount of cores required. Other factors that affect the cost are labor rates, overhead rates, pouring temperature, number of castings per mold, parting line selection, and casting surface finish.

Casting has some distinct advantages over other primary manufacturing processes. Some of these advantages are: Complex 3-D internal and external surfaces can be produced simultaneously; very large products can be produced easily; resulting properties of the products are generally non-directional or isotropic; various alloys can be used; and both low and high levels of production are economical. Some casting processes, such as die casting, are economical only at high levels of production.

7.3 POWDER PROCESSING OVERVIEW

Powder processing is a primary shape forming process. Ceramic products and powder metallurgy are two major processes that utilize powders. Ceramic materials are a major industry, but the procedure is quite different from metals. The emphasis here will be on the process capabilities of the powder metallurgy to compare its capabilities with those of the other metal processing processes. It is a near-net-shape forming process, generally used for small, thin parts. Some of the typical part shapes produced by powder metallurgy are illustrated in Fig. 7.3. There are five steps in the powder metallurgy process:

1. Produce powders
2. Mix powders
3. Form part
4. Sinter part
5. Perform secondary operations

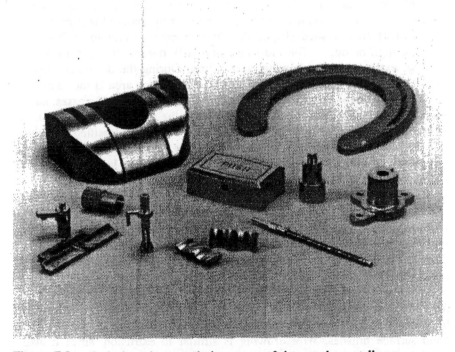

Figure 7.3 Typical products made by means of the powder metallurgy process. (Courtesy of Metal Powder Industries Federation, Princeton, New Jersey.)

There are two main design calculations that frequently occur in the production of powder metallurgy parts. First, it is necessary to determine the amount of materials required to make the part to a specific density before the sintering operation. Second, the change in linear dimension from the volume contraction during sintering must be determined. These two calculations must be considered so that the final part can be manufactured to its appropriate dimensions and specifications.

Powder metallurgy has a high yield compared to other primary processing methods. It is capable of producing complex and thin shapes relatively easily. A part produced via powder metallurgy generally requires few machining operations, because the part usually has a good surface finish. Powder metallurgy also allows for high production levels and good reproducibility.

7.4   BULK DEFORMATION PROCESS OVERVIEW

Bulk deformation, also called metalworking or metalforming, includes processes in which extensive plastic deformation occurs, and the changes in the dimensions of the workpiece are large. The starting material is generally in the form of semifinished shapes that have a high volume-to-surface-area ratio, or high modulus. The most common bulk deformation processes are forging, extrusion, wire drawing, and rolling. During the deformation process, the modulus of the part is reduced as the surface area is increased.

Forging involves applying controlled pressure by use of presses, hammers, dies, or other related machinery to form the material into the desired shape. The process is illustrated in Fig. 7.4. Forging is classified as either hot forging or cold forging, depending on the temperature of the initial workpiece. Forging may be done with an open die, where lateral movement of the metal is unrestricted, or a closed die, where the movement of the metal is controlled on all sides. Impression-die forging contains recesses in the die that allow the workpiece to be formed into complex shapes, and flash is formed. Impression-die forging is usually classified as closed-die forging, but true closed-die forging would have no flash. The production of the images on a coin, such as a dime, is an example of a closed-die forming operation, called "coining."

Extrusion is a plastic deformation process in which metal or plastic is forced to flow through the shaped orifice of a die. The parts generally formed by extrusion are of constant cross-sectional area, with close tolerances and good surface finishes. *Wire drawing* is similar to extrusion, since it also involves forcing the material through a small orifice in a die. However, extrusion generally (but not exclusively) involves pushing the metal through

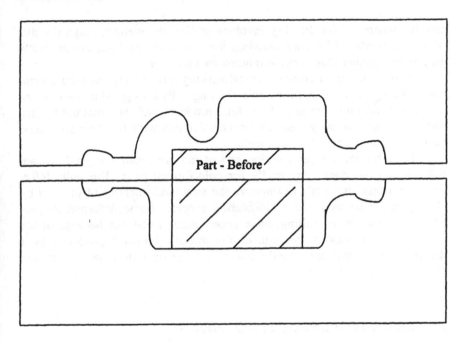

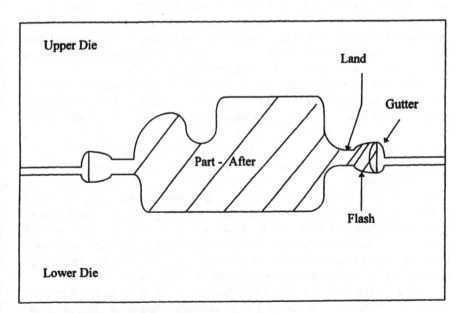

Figure 7.4 Bulk deformation (forging) dies and part.

the die, whereas wire drawing involves pulling the metal through the die. The parts produced by wire drawing, such as wire and bars, are generally longer and thinner than those produced by extrusion.

Rolling, the most common metalworking process, involves the continuous forcing of metal between two rotating rolls on opposite sides of the workpiece. Rolling is used to form final products such as structural beams, and is also used in many cases to form preliminary parts for other processes, such as billets, slabs, and sheet metal.

Parts formed by deformation processing are noted for good surface finishes and tolerances, especially for cold-formed parts. The bulk deformation process also tends to improve the structural integrity of the part by closing up voids from the solidification process. Some deformation processes, such as thread rolling, have good fatigue resistance because of the compressive stresses at the base of the threads. Products produced by a deformation process are usually made in large quantities because of the expensive tooling required.

7.5   SHEET METAL PROCESS OVERVIEW

Sheet metal, or sheet-forming, processes have as the starting material rolled sheet, which has a low volume-to-surface-area ratio, or low modulus. The sheet metal (sheet-forming) processes change the part's shape, but do not greatly alter the value of the modulus. Generally, in sheet metal forming processes, the thickness change of the workpiece is small. Sheet metal processing involving a press is known as *stamping*. The most common sheet metal processes are shearing, drawing, bending, and spinning.

Shearing is used for the cutting of the sheet metal. This procedure is used primarily to produce the initial workpiece, called a *blank*, for another sheet metal process. Shearing is generally done by means of a guillotine or alligator shears. *Drawing* is a process in which a blank is pressed into a shaped die to form an open-ended, cylindrical shape, such as a can. The process is known as *deep drawing* if the cylinder depth is greater than or equal to the radius of the base. The basic components of the deep drawing process, the punch, die, and hold-down ring or pressure plate, are illustrated in Fig. 7.5. *Bending* is a common process in which a straight workpiece is plastically deformed into an angle. This process is, in many cases, part of another sheet metal process. *Spinning* involves pressing a tool into a rotating blank to form hollow shapes such as cylinders, cones, tubes, and hemispheres. It is usually performed on a lathe or similar machine.

Sheet metal processing is used frequently, because most other processes have difficulty in processing thin materials. High production rates are

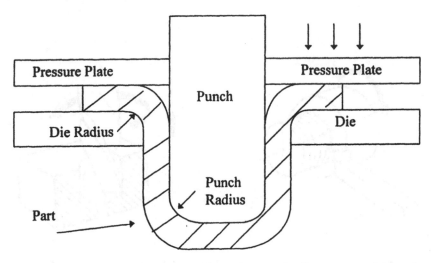

Figure 7.5 Deep drawing process illustrating punch, die, pressure plate, and part.

often attained, and large production quantities are frequently produced. Tooling costs are lower than for bulk deformation processing because the forces required are lower and the processing is usually done at room temperature.

7.6 MACHINING (METAL REMOVAL OR CUTTING) OVERVIEW

Machining involves the shaping of a part through removal of material. A tool constructed of a material harder than the part being formed is forced against the part, causing metal to be cut from it. Machining, also referred to as *cutting, metal cutting,* or *material removal,* is the dominant manufacturing process, because it is the only process used for both primary processing and secondary processing.

Mechanical machining is the most common type of machining process. The basic mechanical machining operations are turning, drilling, milling, and shaping. The basic cutting movements of the cutting operations are illustrated in Fig. 7.6.

1. *Turning*: The turning operation is performed on a lathe where the workpiece rotates and the tool moves parallel to the center axis of the workpiece. The operation is used to produce external cylindrical surfaces for parts such as shafts and axles.

2. *Drilling*: The drilling operation is used to create holes, generally by means of a drill press. The drill press can also be used to improve the surface

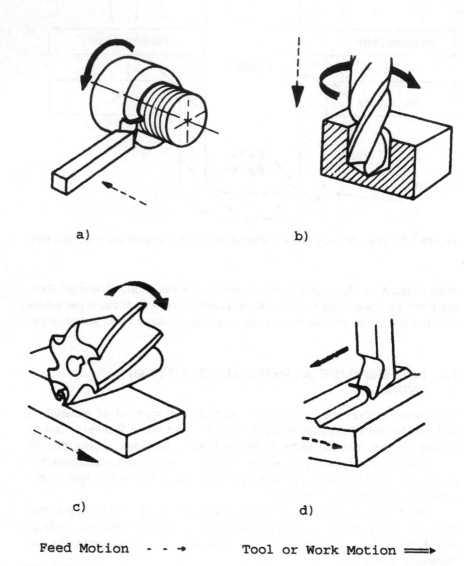

a)   b)

c)   d)

Feed Motion - - →   Tool or Work Motion ⟹

Figure 7.6 Typical cutting movements of tool and work for basic machining processes: (a) turning, (b) drilling, (c) milling, and (d) shaping. (Adapted from Ref. 1 with permission of Marcel Dekker, Inc.)

finish of a hole by reaming, and it can also be used to used to thread a hole, which is known as *tapping*.

3. *Milling*: Milling is an operation using multiple tooth cutters, predominately to generate flat surfaces but also to generate complex surfaces. The milling machine is classified as either a horizontal or a vertical machine. On a vertical milling machine the center axis of the cutter is perpendicular to the cutting table, whereas on a horizontal milling machine the center axis of the cutter is parallel to the cutting table. Milling machines, with the multiple tooth cutters, have high metal removal rates.

4. *Shaping*: The shaper is a relatively simple tool. The workpiece is held in a vice while the ram, which carries the tool, slides back and forth in equal strokes to the desired length of stroke. The tool cuts in one direction only, but the return stroke is faster than the cutting stroke to reduce idle time. The shaper is used mainly for facing, but it can also be used to create slots, steps, and dovetails. The shaper is also used to generate flat surfaces like milling machines, but the metal removal rates are lower because only one tool (or tooth) is used.

Some of the other methods of metal removal are classified with nonmechanical machining. The four basic nonmechanical processes are: (1) chemical milling, (2) electromechanical machining, (3) electric discharge machining, and (4) laser cutting.

Machining is the most widely used manufacturing process, because it can used as a primary shaping process as well as a finishing shaping process. Flat surfaces, cylindrical external surfaces, and holes (cylindrical internal surfaces) are generally formed by machining processes. Machining operations produce good surface finishes and part tolerances and are often used to complete parts whose primary shape was produced by the other processes. The tooling for machining is less costly than for most other processes, and machining can be used for low production quantities as well as high. The production rates are usually lower those of the deformation or casting processes, and operator costs tend to be higher, since production times are longer.

7.7 JOINING PROCESS OVERVIEW

Joining is concerned primarily with the assembly of the components into subassemblies or final products. The joining processes can be divided into three major categories: fasteners, welding, and adhesives. These categories are extremely broad and contain a vast amount of information. Joining focuses primarily on assembly of components or parts into subassemblies or final products.

The three major categories of joining processes follow.

1. *Fasteners*: These are mechanical devices that join materials via clamping forces, pressure, or friction, which do not involve molecular bonding between the surfaces as the primary bonding force. Some examples would be a threaded fastener, pins, and riveting.

2. *Cohesion (welding)*: This involves the joining of two or more pieces of material by means of heat, pressure, or both, with or without a filler metal, to produce a localized union through fusion or recrystallization creating a chemical bond. Welding process are classified by their energy source. The different energy sources used in welding are: electrical, mechanical, chemical, and optical (beam, ray). Some examples of welding processes are: arc welding, resistance spot welding, friction welding, oxyacetylene welding, and electron beam welding.

3. *Adhesion (gluing)*: This is the joining of two or more material components through the forces of attraction between the adhesive and the materials being joined (adherends). Gluing processes depend primarily upon adhesive bonding. Some of the other adhesive processes are brazing, soldering, and epoxy bonding.

Joining processes are used when the structure has a very low modulus and when different material properties are required; for example, the handle of a pot should be a poor conductor, whereas the bottom should be a good conductor, so two different materials are frequently utilized and must be "joined" to form the part.

7.8 SUMMARY

The basic manufacturing processes have been presented to introduce the materials being presented in greater detail in the following chapters. The emphasis is upon the primary shape formation processes and the assembly/joining processes. The chapters will illustrate the major process details and some of the engineering calculations utilized in the process evaluations.

7.9 EVALUATIVE QUESTIONS

1. What are the two main categories of shaping processes?

2. What are the four major areas of the casting process?

3. What are the five steps of the powder metallurgy process?

4. What are the other names for the bulk deformation process?

5. What are the four basic machining operations?

6. What are the two types of milling machines?

7. What are the three major categories of the joining processes?

REFERENCES

1. Alting, Leo. Manufacturing Engineering Processes, 2nd ed., Marcel Dekker, New York, 1994, pp. 205, 208.

INTERNET SOURCES

General manufacturing processes: *http://www.ee.washington.edu/conselc/CE/Kuhn/ manufact/95x2.htm*
Casting processes: *http://www.metalbot.com/cast/html*
Forging: *http://www.quforge.com/*

8
Process Selection Basics

8.1 INTRODUCTION

Process selection is one of the most difficult decisions that occurs in the design and manufacture of a new product. In most current design cases, the only process considered is one that has been used on similar products. If more than one process is considered, only a few alternative processes will be contemplated. When a decision is made about the manufacturing process, most of the product life cycle costs will have been determined. Thus, it is extremely critical that the correct process decision be made to keep costs competitive.

The major difficulty in comparing processes is that the process criteria are different for the major processes. For example, in the casting processes, the process criteria (1) of interest are overall size, section thickness, dimensional tolerances, draft allowance, and machining finish allowance. The process criteria (1) for forgings are thickness, shrinkage and die wear, draft angle, and machining finish allowance. For machined parts, the process criteria included surface finish, rotational symmetry, and L/D ratios. Sheet metal working and powder metallurgy also have different process criteria. Thus, it is extremely difficult to compare processes, and considerable specialized knowledge is required. The traditional, or "seed catalog," approach is to compare the product process requirements with the process performance criteria of different processes in a reference source to find a match. A major problem is that the reference sources often do not list the process performance criteria required by the product.

8.2 PROCESS SELECTION PROCEDURE

A process selection procedure is presented that reduces the problems of the traditional approach and permits new processes to be evaluated by comparison with traditional processes. This procedure eliminates processes from consideration based upon process restrictions, material limitations, and process performance requirements and then selects the best of the remaining processes. A flow diagram of the two-stage process selection procedure is illustrated in Fig. 8.1.

The process selection procedure presented follows that of Edwards and Endean (2) and the Keptner–Tregoe (3) decision process. This is a two-stage decision process, the first stage being the "feasible" stage and the second stage being the "optimal" stage. In the first stage, all the "feasible" characteristics must be satisfied; if not, the process is eliminated from further consideration. In the second stage, all processes that satisfy the first stage are evaluated first upon minimal process criteria and then upon "optimal" process performance characteristics. The remaining process that satisfies best the "optimal" characteristics is selected. In the Keptner–Tregoe decision process, the "must" stage corresponds to the "feasible" stage, and the "want" stage corresponds to the "optimal" stage.

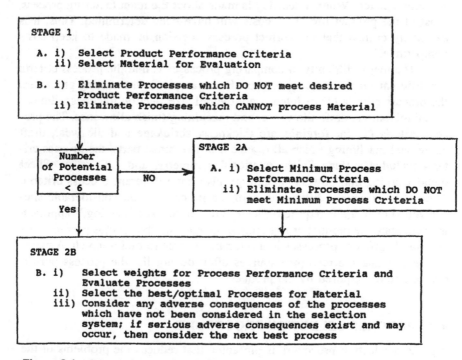

Figure 8.1 Two-stage process selection procedure.

"Feasible" Stage:

Eliminate processes that cannot meet the given design specifications.
Eliminate processes that cannot process the given material.

"Optimal" Stage:

Eliminate processes that cannot meet the minimum acceptable process
performance criteria.
Select the best process from the remaining processes under consideration by optimizing process performance criteria.

8.2.1 Process Feasibility Stage

In the feasible stage, there are two critical criteria: (1) Can the shape be
produced by the process? (2) Can the material be shaped by the process?
Only if these two criteria are met should the process be considered for
further evaluation.

The product criteria for different manufacturing processes are presented in Table 8.1. The descriptions for the criteria and the rating values
for the criteria are presented in Table 8.2. For example, the sand casting
process has a shape rating of 5 for external surfaces. The external shape
rating value of 5 indicates that complex 3D surfaces can be produced by
the sand casting process. The sand casting process has a value of 2 for the
external surface criteria. This indicates that the surface generated by the sand
casting process is a rough surface. This surface may be improved by machining, and the criteria presented are for the surface produced by the process, not for the best surface obtainable, for further operations may improve
the surface. If further processing can be used to improve the surface, then
the surface criteria would not be used as limiting product criteria for the
process.

The material process capabilities for the different processes are presented in Tables 8.3a and 8.3b. These tables indicate the ability of the materials to be shaped by the particular process. For example, cast irons have
a rating of 5 for the sand casting process, which indicates that cast irons
can be shaped easily by sand casting. On the other hand, refractory metals
have a value of 1, which indicates that they are not formed by the sand
casting process. Note that several processes have a value of 5 for cast irons,
whereas only two processes have a value of 5 for refractory metals. Table
8.3a is for metals; Table 8.3b is for nonmetals, such as ceramics, polymers,
and composites. The values in Table 8.3b indicate that only a few of the
traditional processing methods are suitable for processing nonmetals. The
data in Table 8.3b is based on estimates from limited data, and improvement
of the data is needed. However, the primary purpose of Tables 8.1, 8.3a,

Table 8.1 Product Criteria for Different Manufacturing Processes

	Product criteria				
	Shape complexity		Size rating	Surface smoothness	
Primary processes	External	Internal		External	Internal
Casting					
1. Sand casting	5	5	5	2	2
2. Investment casting	5	5	3	5	5
3. Die casting	4	3	2	4	4
4. Permanent mold	4	4	3	4	4
5. Full mold process	5	5	5	2	2
Machining: chip processes					
6. Grinding	5	2	5	5	5
7. Multiple	5	3	5	5	5
8. Single point	5	3	5	5	5
Machining: chipless processes					
9. Electrical discharge	4	2	3	4	4
10. Ultrasonic	2	1	3	3	3
11. Chemical machining	3	2	5	5	5
12. Abrasive jet machining	2	1	4	4	4
Forming: bulk deformation					
13. Forging	5	1	5	3	2
14. Extrusion—standard	3	3	3	4	4
15. Extrusion—impact	3	3	2	5	5
16. Rolling	3	2	5	4	4
Forming: sheet metalworking					
17. Shearing	2	0	5	3	3
18. Bending	3	2	5	5	5
19. Deep drawing	3	3	3	5	5
20. Stretch forming	3	3	3	5	5
Powder processing					
21. Powder metallurgy	4	4	3	4	4
22. Plastic injection	4	3	3	4	4
23. Pressing and firing	4	4	3	5	5
Other processing					
24. Glass drawing/floating	2	2	2	5	5
25. Pultrusion	3	2	2	5	5
26. Blowing	4	4	2	5	5

Secondary processing, such as machining, can greatly improve shape and surface ratings.

Table 8.2 Rating Descriptions for Different Criteria

Rating value	Criteria type 1: shape criteria, external	Criteria type 2: shape criteria, internal	Criteria type 3: size criteria
5	3D complex surfaces	3D complex surfaces	All sizes
4	3D simple surfaces	3D simple surfaces	Volume > 20 cm^3 (large–medium)
3	2½D surfaces	2½D surfaces	Volume < 500 cm^3 (medium–small)
2	2D complex surfaces	2D complex surfaces	Volume < 20 cm^3 (small only)
1	2D simple surfaces	2D simple surfaces	
0	No surfaces	No surfaces	

Rating value	Criteria type 4: shape criteria, external	Criteria type 5: shape criteria, internal	Criteria type 6: material processing difficulty
5	Precision	Precision	Easily processed
4	Smooth	Smooth	Some minor difficulties
3	Medium	Medium	Difficult to process
2	Rough	Rough	Considered possible, but not used
1	Coarse	Coarse	Not considered
0	No surfaces	No surfaces	

and 8.3b is to indicate the type of data needed to evaluate manufacturing processes. The processes must be evaluated by the same product shape criteria, and the criteria used in Table 8.1 are presented as a starting basis.

8.2.1.1 Illustration: "Feasible" Stage

To illustrate the application of the system, an example will be presented. The design engineer has chosen to make the part from aluminum and the product has the following criteria rating values from the design data:

Table 8.3a Material Processing Capability: Metals

Primary processes	Plain carbon	Alloy steels	Cast irons	Light metals	Heavy metals	Refractory metals	Precious metals
Casting							
1. Sand casting	4	4	5	4	4	1	4
2. Investment casting	5	5	2	5	5	5	5
3. Die casting	1	2	1	5	2	1	2
4. Permanent mold	4	4	5	5	5	1	2
5. Full mold process	5	5	5	5	4	1	2
Machining: chip processes							
6. Grinding	5	5	5	4	5	4	5
7. Multiple	5	5	5	5	5	3	5
8. Single point	5	5	5	5	5	3	5
Machining: chipless processes							
9. Electrical discharge	5	5	3	5	5	5	2
10. Ultrasonic	5	5	1	5	5	2	2
11. Chemical machining	5	5	1	5	5	4	5
12. Abrasive jet machining	5	5	5	5	5	5	2

Forming: bulk deformation

13. Forging	5	1	5	3	5
14. Extrusion—standard	5	1	5	3	5
15. Extrusion—impact	1	1	3	1	1
16. Rolling	5	1	5	4	5

Forming: sheet metalworking

17. Shearing	5	2	5	1	5
18. Bending	5	1	5	4	5
19. Deep drawing	5	1	5	4	5
20. Stretch forming	5	1	5	4	5

Powder processing

21. Powder metallurgy	5	5	4	4	5
22. Plastic injection	1	1	3	1	1
23. Pressing and firing	1	1	1	1	1

Other processing

24. Glass drawing/floating	1	1	1	1	1
25. Pultrusion	1	1	1	1	1
26. Blowing	1	1	1	1	1

Table 8.3b Material Processing Capability: Nonmetals

| | Material | | | | | | | | |
| | Ceramics | | | | Thermosetting | Polymers | | Composites | |
Primary processes	Glass	Domestic	Engr.	Electronics		Thermoplastic	Particulate	Fiber	Natural
Casting									
1. Sand casting	1	1	1	1	1	1	1	1	1
2. Investment casting	1	1	1	1	1	1	1	1	1
3. Die casting	1	1	1	1	1	1	1	1	1
4. Permanent mold	1	1	1	1	1	1	1	1	1
5. Full mold process	1	1	1	1	1	1	1	1	1
Machining: chip processes									
6. Grinding	3	4	4	4	4	4	3	3	4
7. Multiple	1	1	1	1	5	5	3	3	4
8. Single point	1	1	1	1	5	5	3	3	4
Machining: chipless processes									
9. Electrical discharge	1	3	3	4	4	4	1	1	1
10. Ultrasonic	4	4	4	4	1	1	4	3	3
11. Chemical machining	3	3	3	3	1	1	1	1	1
12. Abrasive jet machining	4	3	3	3	4	4	4	2	4

Forming: bulk deformation

Process							
13. Forging	1	1	1	1	1	1	1
14. Extrusion—standard	1	1	1	1	1	1	1
15. Extrusion—impact	1	1	1	1	1	1	1
16. Rolling	1	1	1	1	1	1	1
Forming: sheet metalworking							
17. Shearing	1	1	1	1	1	1	1
18. Bending	1	1	1	1	1	1	1
19. Deep drawing	1	1	1	1	1	1	1
20. Stretch forming	1	1	1	3	1	1	1
Powder processing							
21. Powder metallurgy	1	1	1	2	1	1	1
22. Plastic injection	1	1	1	5	1	1	1
23. Pressing and firing	1	5	5	3	5	3	1
Other processing							
24. Glass drawing/floating	5	1	1	1	1	1	1
25. Pultrusion	1	1	1	1	3	5	1
26. Blowing	5	1	1	5	1	1	1

Criteria	Product rating description
Shape criteria, external	2½D
Shape criteria, internal	None
Size rating	18 cm³
Surface criteria, external	Medium
Surface criteria, internal	None

The acceptable process rating values for the shape and surface criteria are all values equal to or greater than the desired rating value. For the size rating, the 18-cm³ size can be in category 2, 3, or 5 (from Table 8.2). That is, 18 cm³ is a small shape size or a medium–small shape size; it is not a large–medium shape size. Processes that can produce only large–medium shapes cannot be considered. The acceptable process rating values for all criteria are as follows:

Criteria	Product rating description	Product rating values	Acceptable process rating values
Shape criteria, external	2½D	3	3, 4, 5
Shape criteria, internal	None	0	All
Size rating	18 cm³	3	2, 3, 5
Surface criteria, external	Medium	3	3, 4, 5
Surface criteria, internal	None	1	All

From Table 8.1 and the product criteria, the following processes can be eliminated from further consideration:

Product criteria	Processes eliminated
External shape	10, 12, 17, 24
Internal shape	None
Size	12
External surface	1, 5
Internal surface	None

The material selected, aluminum, is a light metal, and from Table 8.3a, processes 23, 24, 25, and 26 are eliminated from further consideration. Also, it is known from experience that the only light metal that can use the plastic injection process is magnesium, so this process, number 22, is also eliminated from further consideration. Thus, after the basic shape and material have been determined, 10 of the 26 processes have been eliminated (1, 5, 10, 12, 17, 22, 23, 24, 25, and 26).

8.2.2 Process Performance Evaluation

The second phase of the selection process is to determine the process that meets the product/process requirements in the best, or "optimal," manner. Some of the typical process performance criteria for the second stage of the selection process are cycle time, materials utilization, process flexibility, quality/reliability, operating costs, surface finish, and tolerances. Table 8.4 presents the rating values and descriptions for the various process performance criteria. The critical information is presented in Table 8.5, which contains the process performance rating values for the specific processes. These ratings are approximate and must be further refined for specific applications. The purpose is to indicate how the system can be applied, and data improvement for the system will be a continuous process. The various criteria scales have been developed such that the higher the value, the better or more desirable the specific criteria indicated in Tables 8.2 and 8.4 will be. For example, a complex 3D surface has a rating of 5, an easily processed material has a rating of 5, and a process with excellent quality and reliability has a rating of 5. This rating scale is generally consistent in the evaluation of both the materials and the processes.

8.2.2.1 Illustration: "Optimal" Stage

In the previous illustration, 10 of the 26 processes had been eliminated in the "feasible" stage. A second elimination stage is then recommended, and this is based on the minimum acceptable performance criteria for the product. After the material considerations have been made and the process list is reduced to six or fewer, the optimization stage could begin without a second elimination stage. The second stage required determination of minimum acceptable process performance criteria from Table 8.4. For the product design, the designer would need to select the following minimum process performance criteria:

Table 8.4 Rating Values and Descriptions for Process Performance Criteria

Rating value	Cycle time	Materials utilization	Process flexibility	Quality/reliability R(a) micrometer	Surface finish	Tolerances	Operating costs
5	<20 sec	Small waste, <5 percent of total material	Negligible changeover or setup time (<2 min)	Excellent quality and reliability	<0.80	<0.01 mm	No appreciable setup costs, <$100
4	20 sec to 1 min	Little waste, <15 percent	Fast changeover (2–15 min)	Excellent quality, good reliability	0.80–3.20	0.01–0.05 mm	Low tooling costs, low setup costs, $100–$1,000
3	1–5 min	Waste from 15–40 percent	Average changeover (15–120 min)	Good quality, average reliability	3.20–12.5	0.05–0.25 mm	Medium tooling and setup costs, $1,000–$10,000
2	5–15 min	Waste from 40–60 percent	Slow changeover (120–480 min)	Average quality and reliability	12.5–25	0.25–1.50 mm	High tooling and setup costs, $10,000–$100,000
1	>15 min	>60 percent	Changeover extremely difficult (>480 min)	Poor quality and average reliability	>25	>1.5 mm	Dedicated tooling and high setup costs, >$100,000

Adapted from descriptions in Ref. 2.

Process performance criteria	Minimum acceptable value	Additional processes eliminated by minimum process criteria
Cycle time	3	2, 6, 7, 9, 11, 21
Material utilization	2	6, 7, 8
Process flexibility	2	3, 18
Quality/reliability	3	3, 13, 17, 21
Surface finish	2	—
Tolerances	2	13
Operating costs	2	3, 9, 11, 19, 20

From stage one:

Processes eliminated by material   22, 23, 24, 25, 26
Processes eliminated by shape   1, 5, 10, 12, 17, 24

From stage two:

Processes eliminated by minimum process criteria   2, 3, 6, 7, 8, 9, 11, 13, 18, 19, 20, 21

After these additional processes have been eliminated, the only remaining processes are 4, 14, 15, and 16. These processes meet all the material and shape requirements as well as the minimum process performance criteria. To determine the best process, the one that maximizes the process performance criteria would be selected. If all the criteria have equal weighting, then the sum of the values of the process performance criteria for each process would give a total value and the highest value would be the most desirable process. For the remaining processes:

Number	Process description	Total ranking value
4	Permanent mold	19
14	Extrusion—standard	27
15	Extrusion—impact	27
16	Rolling	25

The two extrusion processes have equivalent ranking values, and thus, weighting of the process criteria should be considered. The rolling process

Table 8.5 Process Performance Criteria Ratings for Specific Processes

Primary processes	Cycle time	Material utilization	Process flexibility	Quality/ reliability	Surface finish	Tolerances	Operating costs
Casting							
1. Sand casting	2	2	5	2	1	1	5
2. Investment casting	2	3	4	4	4	4	3
3. Die casting	5	3	1	1	3	3	1
4. Permanent mold	4	2	2	3	3	3	2
5. Full mold process	1	2	5	2	2	1	5
Machining: chip processes							
6. Grinding	2	1	5	5	5	5	4
7. Multiple	2	1	5	5	4	5	5
8. Single point	3	1	5	5	5	5	4
Machining: chipless processes							
9. Electrical discharge	1	2	4	4	4	5	1
10. Ultrasonic	1	2	3	4	4	3	1
11. Chemical machining	1	1	4	5	5	5	1
12. Abrasive jet machining	4	4	3	2	3	2	4

Forming: bulk deformation							
13. Forging	3	4	3	2	2	1	2
14. Extrusion—standard	5	5	3	3	4	4	3
15. Extrusion—impact	5	5	3	4	4	4	2
16. Rolling	5	4	3	3	4	4	2
Forming: sheet metalworking							
17. Shearing	4	2	3	2	2	2	4
18. Bending	5	5	1	4	3	3	2
19. Deep drawing	5	5	3	4	3	3	1
20. Stretch forming	5	5	3	3	3	2	1
Powder processing							
21. Powder metallurgy	2	5	2	2	2	2	2
22. Plastic injection	5	4	3	3	3	3	2
23. Pressing and firing	1	5	5	3	2	2	1
Other processing							
24. Glass drawing/floating	5	5	4	3	2	2	1
25. Pultrusion	4	3	3	4	3	3	2
26. Blowing	4	4	2	4	3	4	2

is also close in value at this time, but the permanent mold process is much lower in the ranking value.

If it is decided to give different weights to the process performance criteria, such as weight the cycle time and operating costs three times that of the other criteria, the weighted process performance values would be as follows:

Number	Process description	Weighted ranking value
4	Permanent mold	31
14	Extrusion — standard	43
15	Extrusion — impact	41
16	Rolling	39

The weighted ranking value (WRV) for each remaining process is obtained by multiplying the performance criteria weighting factor by the process performance criteria value and summing for each criteria for the process under consideration; that is:

$$WRV = \Sigma \text{ (weighting factor)} \times \text{(process performance criteria value)}$$

For example, for process 4, permanent mold:

$$WRV = (3 \times 4) + (1 \times 2) + (1 \times 2) + (1 \times 3) + (1 \times 3)$$
$$+ (1 \times 3) + (3 \times 2) = 31$$

The process with the highest weighted ranking value would be the most desirable process, which for this problem would be the standard extrusion process. If, however, quality/reliability were selected to have the weighting factor of 3 instead of the operating costs, the impact extrusion process would be selected.

Before the final selection is made, one should consider the possible adverse consequences of the processes under final consideration. The Keptner–Tregoe approach (3) has a process for this, but, in summary, if a process has a potential adverse consequence that has a high probability of occurrence, other processes should be considered. For example, if the only standard extrusion supplier that can make the part is in the process of being purchased by your arch rival competitor, you may want to consider an alternative process.

8.2.3 Secondary Operations

It should be understood that this procedure is for selecting the primary process. In many cases, a primary process may be used to bring the product to "near net shape," and then secondary operations are used to form the final product. For example, sand casting may be used to form the basic shape of a tea kettle, and then secondary operations may be performed, such as drilling and tapping holes for the handle, finishing, and painting. Secondary operations are difficult to address in process selection. Although some computer programs (5) have been developed that include the possibility of secondary operations, more work needs to be done in this area.

8.3 SUMMARY

The selection of what manufacturing process to use to produce a product is very difficult. A process selection scheme is proposed to assist in the selection of the best process. The selection process is a two-stage scheme. The first stage has two parts: (a) eliminate all processes that cannot produce the desired shape; and (b) eliminate all processes that cannot process the desired material. The second stage has two parts: (a) eliminate processes that do not meet the minimum process performance requirements (optional: use when more than six alternatives remain to simplify the decision process); and (b) select the best of the remaining processes based upon the process performance requirements.

The selection of the best process can be based upon equal weighting of all the process performance criteria or the use of higher weights for those performance criteria that are more important. The rating values for the shape criteria, the material processing capabilities, and process performance criteria are for illustrative purposes and are not very accurate. These values were selected to illustrate how the selection scheme can be used and what data needs to be obtained to use the selection process. For actual applications, a more accurate database would need to be developed.

This selection process provides a basis for integrating process selection with material selection and product design. It attempts to evaluate all processes by the same performance criteria so that a valid comparison can be made between widely different processes, such as full mold casting, impact extrusion, and stretch forming.

The selection system presented in this chapter has not been fully developed and needs further development. The chapter also indicates the need for further information about manufacturing processes so products are produced by the "best" process and illustrates the effect of product design upon process selection. These are topics that are presented in the following sections of the book.

8.4 EVALUATIVE QUESTIONS

1. Make a flowchart of the main steps in the process selection procedure.

2. What are some problems with the process selection procedure?

3. Use the following process rating values to answer the listed questions:

		Process rating value		
Process criteria	Weighting importance	Single point cutting	Pressure die casting	Forging
Cycle time	3	2	5	2
Quality	5	5	2	5
Flexibility	2	5	1	2
Material utilization	4	1	4	5
Operating cost	3	5	3	2

a. Which process would be recommended for producing the product?
b. Which process would be recommended for producing the product if all process criteria are weighted equally?

4. Show that if the weighting in the example problem was 3 for quality/reliability and 1 for operating costs, the best process would be the impact extrusion process.

5. a. If a product had the following ratings for the shape criteria, what processes would be eligible for further consideration?

Criteria	Product rating description
Shape criteria, external	3D complex surfaces
Shape criteria, internal	2D complex surfaces
Size	350 cm^3
Surface criteria, external	Medium
Surface criteria, internal	Medium

b. If the material is a refractory metal, what processes remain eligible for consideration?

c. If the internal and external surface criteria had to be precision surfaces instead of medium surfaces, what processes would remain?

6. Which one of the process performance criteria is poorly described? Suggest a better description and rating value for that criterion.

7. a. If a product had the following ratings for the shape criteria, what processes would be eligible for further consideration?

Criteria	Product rating description
Shape criteria, external	3D complex surfaces
Shape criteria, internal	2D complex surfaces
Size	50 cm^3
Surface criteria, external	Smooth
Surface criteria, internal	Rough

b. If the material is copper (heavy metal), what processes remain eligible for consideration if processing is to have only minor difficulties as the worst condition?

c. What processes remain eligible for consideration if the minimum acceptable process performance criteria are as follows?

Cycle time	2
Material utilization	2
Process flexibility	1
Quality/reliability	3
Surface finish	3
Tolerances	3
Operating costs	2

d. If the weighting process is uniform, which is the best of the eligible processes?

8. In the process selection system developed:
   a. What are the product criteria considered?
   b. What are the material criteria considered?
   c. What are the process performance criteria considered?

9. a. If a product had the following ratings for the shape criteria, what processes would be eligible for further consideration?

Criteria	Product rating description
Shape criteria, external	3D simple surfaces
Shape criteria, internal	No surfaces
Size	100 cm^3
Surface criteria, external	Medium
Surface criteria, internal	None

    b. If the material is an alloy steel, what processes remain eligible for consideration if processing is to have only minor difficulties as the worst condition?

    c. What processes remain eligible for consideration if the minimum acceptable process performance criteria are as follows?

Cycle time	3
Material utilization	2
Process flexibility	1
Quality/reliability	3
Surface finish	3
Tolerances	3
Operating costs	2

    d. If the weighting process is uniform, what is the best of the eligible processes?

REFERENCES

1. Dieter, G. Engineering Design: A Materials and Processing Approach, 2nd ed., McGraw-Hill, New York, 1991, pp. 273–369.
2. Edwards, L., and Endean, M. Manufacturing with Materials, Butterworths, Stoneham, MA., 1990.
3. Keptner, C. H., and Tregoe, B. B. The New Rational Manager, Princeton Research Press, Princeton, NJ, 1981.
4. Adithan, M. Modern Machining Methods, S. Chand & Company, New Delhi, 1990.
5. Dargie, P. P., Parmeshwar, K., and Wilson, W. R. D. "MAPS-1: Computer-Aided Design System for Preliminary Material and Manufacturing Process Selection," ASME Transactions, Vol. 104, Jan. 1982, pp. 126–136.
6. Boothroyd, G., Dewhurst, P., and Knight, W. Product Design for Manufacture and Assembly, Marcel Dekker, New York, 1994, pp. 30–61.
7. Alting, L. Manufacturing Engineering Processes, Marcel Dekker, New York, 1994, pp. 127–415.

8. Groover, M. P. Fundamentals of Modern Manufacturing, Prentice-Hall, Englewood Cliffs, NJ, 1996.

9. Ostwald, P. F., and Munoz, J. Manufacturing Processes and Systems, 9th ed., Wiley, New York, 1997.

INTERNET SOURCES

Manufacturing processes: *http://web.staffs.ac.uk/schools/engineering_and_technology/des/aids/process/process.htm*

Expert system for process selection: *http://mml-mac-9.stanford.edu/MMLWebDocs/research/papers/1991/ishii.aieng.91/ishii.aieng.91.html*

9
Metal Casting Process Considerations

Casting is a manufacturing processes with a long and interesting history. Approximately 6500 years ago, copper was forged to make weapons and tools, but it was limited to only a few, relatively simple, shaped tools or weapons. It was not until approximately 500 years later that people in Mesopotamia discovered that a forge fire, operating under perfect conditions, could reduce the copper ore to produce a castable molten product. Intricate shapes of all types could now be made. People began to cast their tools and weapons. The first molds were open molds made in sand. However, people soon discovered that it was more reliable to use more permanent open molds. These molds were cut in stone or limestone or formed in sun-baked clay. This allowed for the repeated production of virtually identical objects. Thus, this was the beginning of the lengthy development of metalcasting. Throughout the many centuries that followed, metalcasting developed into a highly technical and specialized technique, and it is one of the primary manufacturing shaping processes used today (1). Castings had been utilized in the New World before being "discovered"; but once it was discovered, most of the castings were melted into bullion and sent to the "civilized European world." A few actually remain.

Casting generally implies metal casting, but polymers sometimes are referred to as being cast. The major differences are that, in general, castings are poured above the melting point, the liquid has low viscosity and high fluidity, and gravity feeding is common, whereas polymers have high viscosity and low fluidity and cannot be fed by gravity. The remaining discussion will refer to the casting processes; Chapter 10 will consider polymers.

9.1 CASTING OVERVIEW

An overview of the casting process is presented in Figure 9.1, which indicates the four major areas of (1) patternmaking, (2) molding and coremaking,

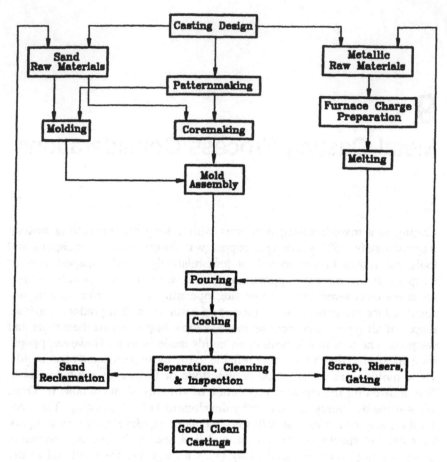

Figure 9.1 Casting process overview.

(3) melting, and (4) cooling and solidification. Patternmaking includes modification of the desired part shape to include allowances for shrinkage, machining, and draft, as well as to provide the gating and risering design. Patterns must also be made for any cores that are used. The primary purpose of cores is to produce the desired internal casting cavities, for example, to make the cooling passages for motor engine blocks. The next phase involves the production of the mold and cores, insertion of the cores in the mold, and the closing of the mold. Simultaneously, while the mold is being prepared, the metal charge (materials that are to be melted to give the proper metal composition) is melted and the molten metal then poured into the mold cavity when the mold is ready. After the metal is solidified, the mold

is separated from the casting and the casting is cleaned and rough finished. The rough finishing includes the removal of the gating system and the risers. Grinding is also done to remove surface imperfections such as fins, which often occur at the parting line.

Like all of the primary manufacturing processes, casting has its own terminology. Some of the commonly used terms follow.

Castability: The ease with which the material responds to foundry practice without special techniques for gating, risering, melting, sand conditioning, or any of the other factors in making a good casting.

Fluidity: The ability of the metal to fill the mold cavity in every detail.

Gating: The metal delivery system in the mold, which includes the pouring basin (if used), the sprue, the runners, and the gates. The sprue, runners, and gates are inside the mold cavity and may be part of the pattern. The pouring basin is on the surface of the mold and can be either part of the mold surface or separate and attached to the mold for pouring. Ceramic filters are often placed in the gating system to prevent loose sand and metal slag or oxides from entering the casting cavity.

Riser (Feeder):   A reservoir of molten metal connected to the casting to provide additional metal to the casting during solidification and to cause directional solidification from the casting to the riser. Its purpose is to eliminate shrinkage cavities in the casting.

Rigging: The engineering design, layout, and fabrication of the pattern to include the location and number of castings, core prints, gating (sprue, runners, and gates), risers, filters, chills, and other special items on the pattern.

Pattern: An image, made of wood, plastic, metal, wax, or styrofoam, used to produce a mold cavity to produce the desired shape of the casting. The pattern is not the same size as the final part, but includes allowances for shrinkage during cooling, for pattern draft to ease removal of the pattern, and for machining on surfaces that must be machined to achieve the desired tolerances or surface finish.

Core: A specially formed material inserted into the mold to form internal surfaces or to form surfaces that cannot be shaped easily by the pattern. The core print is that portion of the mold pattern that supports the core in the mold. The cores must be removed after the casting has solidified and must be designed so they can be removed relatively easily.

Mold: The form that contains the cavity into which the metal is poured to produce the desired shape. The mold can be made of sand (with various binders or vacuum), metal, or refractory material. Cores are inserted into the mold to form internal surfaces of the part, whereas the mold forms the external surface of the part.

The casting processes can be classified by the type of mold used (permanent or expendable/consumable), by the type of pattern, which also can be either permanent or expendable/consumable, and by the type of production (either continuous or discrete). Processes that have permanent molds generally require high production quantities because of the high expense of the mold. The consumable molds can be either high or low production. The patterns for consumable molds may be either permanent patterns, which can be reused, or expendable, as in the lost-wax process or the styrofoam process. Expendable patterns tend to be higher cost, but they permit production of more intricate shapes. Permanent patterns are made of wood, plastic, or metal, depending upon the "permanence" required of the pattern. A classification system of the casting processes based upon their mold and pattern or production type is presented in Table 9.1.

Most castings produced are of the consumable mold and permanent pattern type. Within this category, the most common type is green sand casting. Green sand is not green in color, but rather has water in it. The water is used to activate the clay and make it "sticky" to bind the sand grains together. These are the three main ingredients of the green sand system: sand, clay, and water. Other additions are made to improve the surface finish of the casting. In sand molding, the sand is often 3–10 times the weight of the casting, and recycling or resuse of the mold material is essential. The sand commonly used is silica sand, but other sands used are olivine, chromite, and zircon. These other sands are more expensive and have higher densities than silica sand, but they do give higher cooling rates and are more environmentally acceptable. In some parts of the world, silica sand is considered as a carcinogen and it can cause silicosis. However, the United States has had no reported cases of silicosis attributable to foundry operations in over 20 years, since numerous safety precautions have been taken to prevent silicosis. The typical binder is clay, and commonly used clays are the bentonites (sodium and calcium) and kaolinite.

Other types of binders and additives are used for a variety of reasons, but the main reasons are better surface finish, faster production times, lower labor skills required, and lower production costs. Finishing operations in the foundry are generally labor intensive and thus very costly, so processes that improve the surface finish can greatly reduce production costs and the unit production time.

The process yield is typically low, that is, 30–60 percent of the total metal poured. The gating and risering systems used to produce the casting

Table 9.1 Classification System for Casting Processes Based upon Mold, Pattern, and Production Type

Permanent mold		Consumable mold	
Continuous production	Discrete production	Permanent pattern	Consumable pattern
Continuous casting	Permanent mold (gravity)	Bonded sand casting Green sand Chemical bonding Sodium silicate Shell Organic no-bakes Furan Isocyanate	Investment casting (wax)
	Die casting (high pressure)	Vacuum casting V-Process (unbonded sand)	
	Rheocasting (slush casting) Low-pressure die casting	Plaster/ceramic mold Centrifugal casting Sand molds	
	Centrifugal casting Metal Molds (pipe castings)		Styrofoam casting (full mold) (evaporative casting)

can account for a large portion of the metal poured, especially on small castings. The process yield, commonly called *yield*, is defined as:

$$\text{yield (\%)} = \text{(Casting Weight)/(Total Weight Poured)} \times 100 \qquad (9.1)$$

where

$$\text{total weight poured} = \text{casting} + \text{gating} + \text{risers} + \text{melt loss} \qquad (9.2)$$

The melt loss occurs from slag or dross formed and from volatilization of some metals, such as zinc in copper alloys. It is usually in the range of 2–5 percent of the metal poured, but may be as high as 10 percent.

There are various types of casting processes in use today. A brief description (2,3) of a few of the more common casting processes is presented, but more details can be found using the Internet sources at the end of the chapter.

1. *Sand casting*:   This technique involves consumable molds made of sand that contain varying amounts of resin binders. The most common type of sand casting is green sand casting. Gravity is utilized to feed molten metal into the mold (4). The process is most widely used for iron and steel castings because of their relatively high melting points, which restrict the use of some of the other casting processes. Sand casting is also commonly used for copper-base and aluminum alloys. The sand casting process is applicable for parts ranging in weight from a few pounds to several tons. With sand casting, parts of great complexity and moderate dimensional accuracy can readily be produced. Figure 9.2 illustrates the basic sand casting process using a pattern, core, and cope and drag mold.

2. *Permanent mold casting*:   This process uses a permanent mold constructed of two halves hinged or clamped to each other. This process is generally limited to casting aluminum and magnesium parts, although cast iron, ductile iron, zinc, copper, stainless steel, and lead sometimes utilize the permanent mold process. The permanent mold casting process is illustrated in Fig. 9.3a along with some typical part shapes.

3. *Investment casting*:   This process utilizes consumable ceramic molds and either wax or plastic patterns. These molds are preheated in a furnace to remove the pattern by melting it and letting it drain from the mold. After the wax is removed, the mold cavity will be filled with molten metal. Almost any metal can be cast using investment casting. Metals that have melting temperatures that are too high for plaster or metal molds can be cast using this process. There is almost no limit to the complexity of the shapes that can be produced using investment casting. The weight of typical castings ranges from a few ounces to 40 lb. (4). One of the molds used is illustrated in Fig. 9.3b for the investment casting process along with some typical intricate shapes produced by the process.

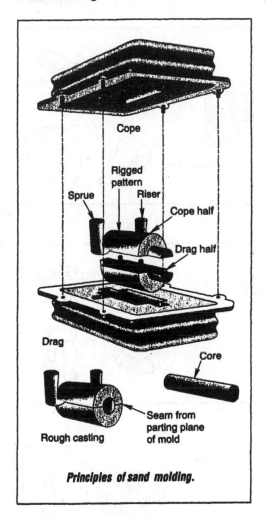

Principles of sand molding.

Figure 9.2 Basic components in sand casting process: Pattern, Core, Cope, and Drag. (Reprinted with permission of Penton Publishing Co. from Ref. 3.)

4. *Die casting*: This process is often considered the best alternative for high production rates. Molten metal, usually zinc or aluminum, is forced under high pressure into closed metallic dies. The complexity of the parts that can be produced using die casting is more limited than with sand or investment casting, since the dies are solid. The cold chamber die casting process is illustrated in Fig. 9.3c along with some of the more complex die casting products.

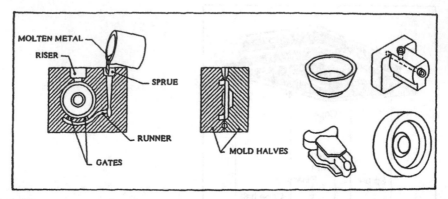

(a)

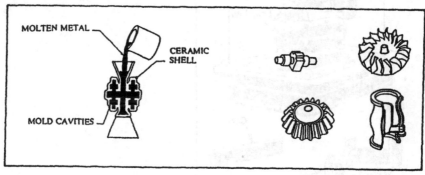

(b)

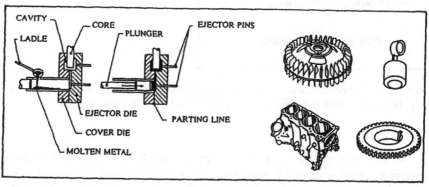

(c)

Figure 9.3 Permanent mold, investment casting, and die casting processes with typical products. (a) Permanent mold casting and typical products. (b) Investment casting and typical products. (c) Cold chamber die casting and typical products. (Reprinted from Ref. 11 with permission of Marcel Dekker, Inc.)

5. *Plaster mold casting*:   This technique utilizes molds made from a slurry of gypsum. The castings have fine finishes and molding detail comparable to investment castings. Plaster mold casting produces parts that are usually small and weigh no more than 20 lb. This process is suitable for thin-wall parts, usually as thin as $1/16$ inch. Copper alloys are the most common materials cast in plaster molds. However, aluminum and magnesium are also commonly cast by the plaster mold process.

The advantages for selecting casting over the other primary processes follow:

1. *Complications of shape*:   The casting method is the only method that can produce complex, three-dimensional, internal and external surfaces simultaneously. Frequently, it is the only method for producing parts of intricate shapes, such as automotive engine blocks, as a single piece.

2. *Product size*:   Very large products are difficult, if not impossible, to make by other processes. Diesel locomotive engines can approach 1000 ft^3 in volume (over 30 m^3), and the other primary processes cannot produce such large, bulky objects with complex internal surfaces.

3. *Resulting properties*:   Resulting properties of the product are generally nondirectional or isotropic, whereas wrought products generally have directional or anisotropic properties.

4. *Alloy properties*:   The alloy may be suitable only for casting. Cast iron, for example, cannot be forged. Refractory metals, such as vanadium, chromium, and tungsten, are difficult to machine.

5. *Economics*:   The economics of casting are viable at low production levels, such as one or two units, or at high levels for complex sand castings, such as engine blocks, manifolds, or water meter housings, and for permanent mold processes. The tooling used for low production is "soft" tooling, that is, wax, wood, or plastic; whereas for high production the tooling typically is "hard," that is, metal patterns.

9.2 DESIGN CONSIDERATIONS IN CASTING

The casting process represents the most economical process of going from raw material to finished product. In general, casting should be the first choice for the primary process selection. The other processes should be selected only when the casting process cannot meet the desired design requirements in a competitive manner. With the current era's emphasis on "near net shape" manufacturing, casting is the primary manufacturing process best suited to near-net-shape manufacturing. New casting processes and new techniques are being developed, so the metal casting industry remains competitive. New materials, such as the metal matrix composite materials, frequently require new casting processes to be developed to make the processes "economically" feasible.

The major design in casting occurs in the tooling area, that is, in the patternmaking for the pattern and in the core box design for the cores (if cores are used). The pattern is not the same shape or size as the desired product. The casting will shrink as it cools from the solidification temperature to room temperature via thermal contraction. Thus, the pattern must be larger than the final product to allow shrinkage to the desired size. Also, in order to separate the pattern from the mold, some draft may be required, and this must be added to the pattern shape. Other allowances are added for machined surfaces, because a certain minimum of material must be removed. Design modifications to the pattern will also occur from the gating and risering systems required to control the metal flow and solidification. Since the cores are a solid mass and form the internal surfaces, little shrinkage will occur.

The two areas of design of major emphasis in casting that are not considered in the other primary shaping processes are gating design and riser design. The casting process involves the pouring of liquid metal into a mold to solidify to the desired shape. The gating system involves the metal pathway from the pouring basin or opening in the mold to the mold cavity that forms the shape of the product. The gating system design objectives are to get the metal into the mold cavity as quickly as possible without eroding the mold, dissolving gases, forming oxides, or causing other harmful effects. Gating design considerations include factors such as where the metal should enter the mold cavity; the size and shape of the sprue, runners, and gates to control the metal flow into the mold cavity; and the use of filters to clean the metal.

During the solidification process, the metal will shrink, and additional metal must be provided to account for the shrinkage. Risers, or feeders as they are called in the European literature, can be used to provide feed metal to account for the shrinkage due to the liquid contraction as the metal cools from the pouring temperature to the solidification temperature and due to the contraction that occurs as most metals go from the liquid to the solid state. The risers must be located so that directional solidification will occur from the casting to the riser, and they must be sized properly so they will have sufficient feed metal to account for the shrinkage that occurs. The risers must be designed so that they will solidify after the casting, provide sufficient feed metal, and be located so that directional solidification will occur. More details on gating and riser design are in the following sections.

9.3 CASTING QUALITY CONSIDERATIONS

The casting process involves two major steps, molding (including coremaking and the setting of the cores) and melting (including the pouring of the metal). Casting defects can occur because of problems in either step or as a result of marginal conditions in both areas. Thus, it is difficult to track down the cause of defects when they occur at a small rate. For example, gas porosity in the casting can be caused by excess gas dissolved in the molten metal or from moist surfaces at or near the mold surface. The foundry industry has classified seven major types of defects (5):

1. *Metallic projections*: fins, flash, swells, and other projections
2. *Cavities*: blowholes, pinholes, dispersed shrinkage, micro- or macroshrinkage, centerline shrinkage, core shrinkage, corner shrinkage, leakers
3. *Discontinuities*: cracks, hot or cold tears, cold shut
4. *Surface irregularities*: seams, rough surface, grooves on surface, depressions on surface, inclusions on surface, burn on, burn in, etc.
5. *Missing portion of casting*: misrun, pour-out, fractured casting, etc.
6. *Incorrect dimensions or shape*: improper shrinkage allowance, incorrect pattern, irregular contraction
7. *Inclusions*: metallic inclusions; slag, dross, or flux inclusions; mold or core materials

Process control in patternmaking, molding, and coremaking and in melting and pouring must occur to prevent most of these defects. Fracture can occur in the shakeout if the casting is not cooled sufficiently. The quality problems are often complex, because inclusions can be from the cores, from the mold, from the slag during pouring, or from a variety of other factors. Finding the specific cause of the defect can be difficult. After the cause of the defect has been determined, one must develop a remedy and a procedure to test the remedy, as well as a monitoring procedure to ensure that the defect cause and defect do not occur again and to see that the proper documentation of the defect, defect cause, and remedy is maintained in case the defect occurs in other products.

9.4 BASIC THEORETICAL FUNDAMENTALS FOR CASTING

The theory of casting involves nucleation and growth fundamentals, heat transfer fundamentals, fluid flow, and stress analysis at high temperatures. Modeling of the casting process initially focused upon heat transfer and solidification shrinkage, and new state-of-the-art efforts are being made on fluid flow analysis, microstructure prediction, mechanical property prediction, and residual stress analysis. New solidification models are being developed using various criterion functions to predict casting porosity and dendrite arm spacing.

Basic nucleation theory considers two types of nucleation: heterogeneous nucleation and homogeneous nucleation. *Homogeneous nucleation* considers the formation of a spherical drop and then how the drop will grow. Most of the theory developed deals with homogeneous nucleation, but the type that generally occurs in casting is heterogeneous nucleation. *Heterogeneous nucleation* includes the effect of special nucleation sites, such as the mold walls or dendrites that have broken off during the solidification process and are in the liquid. Surface tension is important in the formation of the spherical cap at the surface, and this is affected by the mold material as well as by the metal.

The types of grains formed in casting are generally classified as columnar grains or equiaxed grains. A *grain* is a stable nucleus that forms and grows during solidification. *Columnar grains* grow in a preferred direction, usually perpendicular from the cooling edge toward the thermal center of the casting. Equiaxed grains are formed when there is a high degree of undercooking in the liquid, and they tend to grow in all directions rather than one preferred direction. In most cases, equiaxed grains are preferred, because the material has higher strength and is less brittle. The two types of grains are illustrated in Fig. 9.4. At high temperatures, the grain boundaries increase the creep behavior of the material, which is harmful. For the production of turbine blades, single grains, called *single crystals*, are used, because they have been solidified under controlled conditions so that the

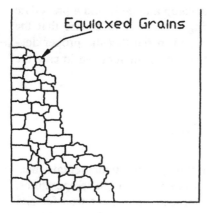
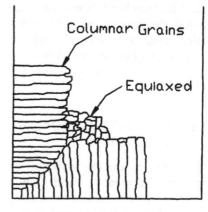

Equiaxed Structure   Columnar Structure

Figure 9.4   Equiaxed and columnar grains.

blade is a single grain. The solidification process is extremely slow, and the blades are more expensive than blades that have equiaxed grains or columnar grains, but the improved performance is worth the additional cost and reduces the total life cycle cost.

The heat transfer mechanism that controls the solidification varies with the different casting processes. The heat transfer can be considered to occur from the liquid metal to the solidified metal to the mold material. The mold material and the interface between the mold material and solidified metal greatly affect the solidification process. Three different cases of heat transfer are generally considered:

1. Insulating mold control
2. Mold–metal interface control
3. Conducting mold control

The insulating mold control is the most general case and is applicable in most casting processes. In this case, the main temperature drop occurs in the mold. The time for solidification can be expressed by Chvorinov's rule:

$$t = C(V/SA)^2 \qquad (9.3)$$

where

t = solidification time
C = constant based upon mold material properties, solidification temperature, and pouring temperature
V = casting volume
SA = casting cooling surface area

The ratio V/SA is called the *casting modulus*, and Chvorinov's rule is often written as:

$$t = CM^2 \qquad (9.4)$$

where

$$M = V/SA \qquad (9.5)$$

The second case, where the interface between the mold and the solidified metal is controlling, occurs in gravity die casting, commonly called *permanent mold casting*, because the solidified metal pulls away from the mold cavity and an air gap forms. The temperature drop occurs across the gap rather than in the mold or solidified metal. The heat transfer across the gap is convection and radiation rather than conduction. The exponent of Chvorinov's rule changes from 2 to 1 and the expression for C also changes.

In the third case, in which conducting molds are used, the major temperature drop occurs in the solidified metal. This heat transfer model is used for the die casting process, also called the *pressure die casting process*, and in plastic injection molding. Chvorinov's rule seems to apply to conducting molds, in that the exponent is 2, but the expression for C is different than that used for the insulating mold case.

9.5 FLUIDITY CONSIDERATIONS

The fluidity of a metal, that is, the ability to fill the mold and reproduce the mold detail, is affected by the mold, the metal composition, and the metal pouring temperature. Molds with insulating properties will improve fluidity, because less heat is removed by the mold. Better surface finish improves fluidity, because rough surfaces slow the flow of the metal. The formation of surfaces films between the metal and mold should be avoided.

Metals with single freezing points or narrow freezing ranges generally have the best fluidity. Compositions with wide freezing ranges should be avoided. High metal velocities, caused by high-pressure heads, generally increase fluidity, which is desirable. However, these high metal velocities also increase turbulence, gas porosity, mold erosion, and several other undesirable effects.

The fluidity of the metal increases linearly with the amount of superheat for up to 200°C of superheat. *Superheat* is the degrees of temperature above the liquidus temperature of alloy. The *freezing range* is the temperature range between the liquidus and solidus temperatures. A narrow freezing range is generally less than 50°C, whereas a wide freezing range is greater than 150°C. Although superheat improves fluidity, it also increases the tendency for gas porosity and inclusions from oxide films; thus, excess superheat is to be avoided.

A secondary factor for consideration in fluidity is the back pressure caused by the air in the mold as the metal enters the mold cavity. If the mold has low permeability, back pressure will develop that will retard the flow of the metal into the mold and cause an apparent loss of fluidity. It can also cause "misruns," an incomplete filling of the mold cavity.

The two tests commonly used to measure fluidity are the spiral fluidity test and the vacuum fluidity test. The *spiral test* is the test commonly used in the foundry, because it includes the effect of the mold materials. The *vacuum test*, also called the *suction tube test*, is more indicative of the metal viscosity rather than of the metal fluidity. These two tests are indicated in Fig. 9.5.

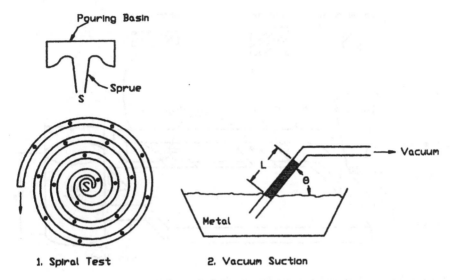

Figure 9.5 Measures of fluidity in metals.

The spiral test measures the length of the metal spiral created when the metal is poured into the mold produced with the standard spiral test pattern. As the superheat increases, the spiral length increases approximately 5 cm/10°C (0.1 in./°F) superheat increase, that is, a 0.5-cm length increase per degree centigrade superheat. The vacuum test measures the length that the metal is drawn into the tube; the longer the length, the better the fluidity.

9.6 BASIC GATING AND RISERING DESIGN CALCULATIONS

The design calculations for gating and riser design presented are the preliminary or basic designs. Further modification of the equations would result from the application of a higher level of fluid flow or heat transfer than the basic level assumed. More advanced relationships can be found in the Bibliography at the end of the chapter. Only the basic formulas and design rules for gating and risering are presented here. A more advanced approach can be found in Ref. 6.

9.6.1 Basic Gating Relationships

The casting and gating system is shown in Fig. 9.6; it indicates the sprue, runner, and gates. The *sprue* is where the metal enters the mold cavity; for large castings there may be more than one sprue. The sprue is also referred

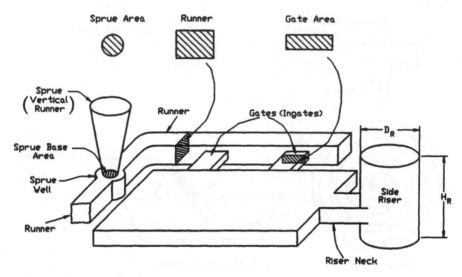

Figure 9.6 Gating and risering for simple plate casting.

to as the *vertical runner*, for it transports the metal from the surface to the plane of the casting. The runners transport the metal from the sprue to the various gates that are connected to the casting. *Gates* control the metal flow from the runner to the casting. The gates are rather short in length, 5 cm (2 in.), whereas the total runner length approaches the perimeter of the casting. The length of the sprue is controlled by the flask size, and the length (or, more appropriately, height) of the sprue is generally determined as a function of the cope height. The basic gating relationships are founded on the first laws of fluids, such as the law of continuity and of the conservation of energy. Frictional losses are not considered in the basic relationships presented here, but they are included in Refs. 6 and 7.

The major function of the gating system is to deliver clean metal into the mold cavity without deteriorating the quality of the metal. This includes items such as:

1. Avoiding the passage of slag or dross into the casting
2. Avoiding the production of dross during passage through the gating system into the casting
3. Minimizing the entrainment of air or mold gases into the metal stream
4. Avoiding mold and core erosion
5. Introducing metal at a rate to minimize shrinkage and distortion

Secondary considerations include:

1. Ease of molding of the gating system
2. Ease of removal of the gating system from the casting
3. Improvement of the casting yield

One of the terms used in gating is the *gating ratio*, which is the ratios of the total sprue base area to the total runner cross-sectional area to the total gate cross-sectional area, shown as:

$$A_S:A_R:A_G \tag{9.6}$$

where

A_S = total sprue base cross-sectional area
A_R = total runner cross-sectional area of all runners
A_G = total gate cross-sectional area of all gates

These different cross-sectional areas are illustrated in Fig. 9.6. This relationship is based upon the continuity of flow, that the same volume of metal must pass through the sprue, then through the runners, and finally through the gates into the mold cavity. That is:

$$Q_S = Q_R = Q_G \tag{9.7}$$

or

$$A_S V_S = A_R V_R = A_G V_G \tag{9.8}$$

where

Q = flow rate (volume/time)
A = cross-sectional area
V = velocity (length/time)

The two types of gating systems referred to are the pressurized gating system and the nonpressurized system. In the *pressurized gating system*, the condition is:

$$A_G < A_S$$

which, if the flow rate is constant, implies:

$$V_G > V_S$$

The *choke* of a system is defined as that section with the lowest total cross-sectional area. In the pressurized system, the gate is the choke and thus the velocity is highest as the metal enters the mold cavity.

In the *nonpressurized system*, the conditions are reversed and the sprue becomes the choke; that is:

$$A_S < A_G \quad \text{or} \quad A_g > A_S$$

and correspondingly,

$$V_S > V_G \quad \text{or} \quad V_G < V_S$$

Most metals generally have nonpressurized gating systems, but some steel castings have slightly pressurized systems. If frictional losses are considered, some nonpressurized systems may behave as pressurized systems and fill the gates. A full gating system will reduce the aspiration of gases into the metal from the vena contracta, but the high velocities can result in turbulent flow that will increase the gas content of the metal.

The first step in gating design is to determine the choke velocity. The choke may be either the sprue base or the gate; it generally is never the runner. From the conservation of energy, the choke velocity can be determined by:

$$V_{choke} = \sqrt{2gh} \tag{9.9}$$

where

g = acceleration of gravity
h = effective sprue height

Campbell (6) has suggested that restrictions should be placed on the velocity of metals in the gating systems. He recommends that the velocity at the gates be restricted to 150 mm/sec (6 in./sec) for heavy alloys (ferrous and copper alloys) and 500 mm/sec (20 in./sec) for light metals (aluminum). These limits are relatively new (1991) and should be used as design restrictions.

The flow rate can be determined from the volume of metal poured (casting plus gating plus risers) and the pouring time. The pouring time is usually specified for the casting; thus the flow rate is:

$$Q = \text{Vol}/t \tag{9.10}$$

where

Q = flow rate (volume/time)
Vol = total volume poured
t = pouring time

The total volume poured can be approximated by taking the casting weight divided by the estimated metal yield, since the actual gating and risering system has not be determined at this stage.

The area of the choke can be found from the flow rate and the choke velocity as:

$$A_{choke} = Q/V_{choke} \tag{9.11}$$

$$= \text{Vol}/(t\sqrt{2gh} \tag{9.12}$$

Once the choke area is determined, the remaining areas can be determined from the gating ratio. The lengths of the runners and gates are determined by the shape of the casting. The total volume poured must be recalculated after the gating and riser systems have been designed, to verify that the total volume poured is near the value used for the choke area calculation. The recalculation will not be performed in the example problems presented.

The Reynolds number is used to determine whether the flow is laminar or turbulent. In casting, turbulent or severely turbulent flow is to be avoided. Turbulent flow starts when the Reynolds number exceeds a value of approximately 2000. Severely turbulent flow, for metal casting, is when the turbulence causes a breaking of the laminar surface flow and permits gases to dissolve in the metal. The Reynolds number for severely turbulent flow is nearly 20,000 for metals such as aluminum, whereas for other metals, such as ductile iron, the Reynolds number for severely turbulent flow is 2000. The Reynolds number equation is:

$$R\# = VDd/\mu \tag{9.13}$$

where

$R\#$ = Reynolds number (dimensionless)
V = velocity (cm/sec)
D = flow channel diameter (cm)
d = density (gm/cm^3)
μ = dynamic viscosity (poise = g/cm-sec) (centipoise = 0.01 poise)

The flow channels for the runners and gates are noncircular in crosssection, so an expression to determine an equivalent diameter is:

$$D(eq) = 4A/P \tag{9.14}$$

where

$D = D(eq)$ = equivalent flow channel diameter (cm)
A = channel cross-sectional area (cm^2)
P = channel cross-sectional perimeter (cm)

The Reynolds number for metals in casting is generally quite high, and turbulent flow is usually obtained. However, the flow is sometimes controlled to avoid the severely turbulent conditions. Other dimensionless numbers, such as the Weber number, are sometimes used for flow control in castings (6).

   Example Problem 9.1.   A steel plate casting 1 in. × 4 in. × 12 in. is poured in 10 sec, the effective sprue height is 4 in., and the gating ratio is 1:2:3. The density of steel is 7.8 g/cm^3 (0.28 lb/in.^{3}) and the casting yield is 60 percent. The cylindrical tapered sprue is connected to two square runners, and each runner is connected to two gates that have a width three times the height. The dynamic viscosity of steel is 6 centipoise.
   Determine the following:

   a.   The amount of metal poured (lb and in.^{3})
   b.   The pouring rate (lb/sec and in.^{3}/sec)
   c.   The choke velocity (in./sec)
   d.   The choke area and the location of the choke
   e.   The dimensions of the sprue base, each runner, and each gate (in.)
   f.   The maximum Reynolds number

Solution:

   a.   The volume of the casting is 1 × 4 × 12 = 48 in.^{3} So:

   volume poured = casting volume/casting yield

$$= 48 \text{ in.}^3/0.60 = 80 \text{ in.}^3$$

   weight poured = volume × density

$$= 80 \text{ in.}^3 \times 0.28 \text{ lb/in.}^3$$

$$= 22.4 \text{ lb}$$

   b.   flow rate = pouring rate/density

$$Q = \text{vol}/t = 80 \text{ in.}^3/10 \text{ sec} = 8 \text{ in.}^3/\text{sec}$$

$$P = \text{pouring rate} = Q \times d$$

$$= 8 \text{ in.}^3/\text{sec} \times 0.28 \text{ lb/in.}^3$$

$$= 2.24 \text{ lb/sec}$$

   Also,

$$P = 22.4 \text{ lb}/10 \text{ sec} = 2.24 \text{ lb/sec}$$

   c.   choke velocity = $V = \sqrt{2gh}$

$$= \sqrt{2 \times 32.2 \text{ ft/sec}^2 \times 12 \text{ in./ft} \times 4 \text{ in.}}$$

$$= 55.6 \text{ in./sec}$$

   Note: This velocity is much higher than the recommended 6-in./sec gate velocity recommended by Campbell for heavy metals.

d. If the gating ratio is 1:2:3, the choke is the sprue base, because it has the smallest value. The velocity at the gates would be 1/3 the velocity of the choke, so the gate velocity would be 18.5 in./ sec, which is still 3 times the recommended value for heavy metals. Friction from the walls of the gating system would slightly reduce the metal velocity. Consideration would be given to using a ceramic filter in the runner, which, in addition to cleaning the metal, would greatly reduce the metal velocity. This, however, is only an example problem.

The choke area can be determined from the flow rate and the choke velocity by using Eq. (9.11):

$$A_{choke} = Q/V_{choke}$$
$$= 8 \text{ in.}^3/55.6 \text{ in./sec}$$
$$= 0.144 \text{ in.}^2 \tag{9.11}$$

e. The sprue base is the choke area, and since it is a tapered cylinder, the cross-sectional area is given by $\pi D^2/4$. Solving for D, one obtains:

$$D = \sqrt{(4 \times A_{sprue\,base})/\pi}$$
$$= 0.428 \text{ in.}$$
$$= 0.43 \text{ in.} \tag{9.15}$$

There are two runners; with a square cross section, the area can be represented as X^2. The total runner area is twice the sprue base area from the gating ratio, so:

2 runners $\times$ X^2/runner $= 2 \times 0.144 \text{ in.}^2$

Solving for X, one obtains $X = 0.38$ in.

There is a total of four gates, with the width being three times the height, and the total gate area is three times the choke area from the gating ratio. If one lets H be the height of the gate, then using the area values:

$4 \times (H \times 3H) = 3 \times 0.144 \text{ in.}^2$

Solving for H, one obtains, $H = 0.19$ in. And $W = 3H = 0.57$ in.

f. The maximum Reynolds number will occur at the location where the maximum velocity occurs for a single channel. The maximum velocity occurs at the choke, and thus:

$R\# = VDd/\mu$

$$= \frac{55.6 \text{ in./sec} \times 2.54 \text{ cm/in.} \times 0.43 \text{ in.} \times 2.54 \text{ cm/in.} \times 7.8 \text{ g/cm}^3}{6 \text{ centipoise} \times 0.01 \text{ poise/centipoise} \times 1 \text{ g/cm-sec/poise}}$$

$= 20{,}052 = 20{,}000$

The Reynolds number at the gates requires a calculation of the equivalent flow channel diameter:

$D(\text{eq}) = 4A/P = 4 \times (0.19 \times 0.57)/[2 \times (0.19 + 0.57)]$

$= 0.285$

The velocity at the gates (18.5 in./sec) is one-third that of the choke, since the cross-sectional area is three times as large, so the Reynolds number at the gate would be:

$$R\# = \frac{18.5 \text{ in./sec} \times 2.54 \text{ cm/in.} \times 0.285 \text{ in.} \times 2.54 \text{ cm/in.} \times 7.8 \text{ g/cm}^3}{6 \text{ centipoise} \times 0.01 \text{ poise/centipoise} \times 1 \text{ g/cm-sec/poise}}$$

$= 4{,}422 = 4{,}400$

If the gate velocity were reduced to the 6 in./sec value as recommended by Campbell (6), the Reynolds number would be less than 2000 and flow would be laminar in the gates. As previously mentioned, the use of a filter at the sprue base or in the runner is one method to reduce the metal velocities in the gates and into the casting cavity.

9.6.2 Basic Riser Design

Risers, or feeders as they are called in the European literature, are used to supply liquid metal to the casting to prevent shrinkage from the thermal contraction of liquid metal as it cools and from the liquid-to-solid transformation. The riser cannot provide for the contraction in the solid state; this must be done in the design of the pattern.

There are six design rules (6) that should be considered when designing risers:

1. *Heat transfer requirement.* The riser must solidify after the casting.
2. *Feed metal requirement.* The riser must have sufficient feed metal for the casting.
3. *Feeding path requirements.* The riser must be located so directional solidification will occur, a minimum temperature gradient exists in the casting, and the feeding distance is less than the maximum.

4. *Junction requirement.* The junction between the riser and the casting should not create a hot spot. A hot spot occurs if the junction has a higher solidification time than the riser or the casting.
5. *Pressure gradient requirement.* There must be sufficient pressure differential to cause the feed metal to flow.
6. *Pressure requirement.* There must be sufficient pressure to prevent the formation and growth of cavities.

Although all six requirements are important, the emphasis will be placed upon the first three requirements, because design relationships can be expressed mathematically for these requirements. The heat transfer requirement is based upon Chvorinov's rule and ensures that the time of solidification of the riser is greater than the solidification time for the casting. The two types of risers considered will be the top and the side cylindrical risers. The side riser is illustrated in Fig. 9.6. The basic design rules developed (8) are:

Top riser: $H = D/2$ $\qquad\qquad\qquad\qquad$ (9.16)

Side riser: $H = D$ $\qquad\qquad\qquad\qquad$ (9.17)

and for both:

$$D = 6M_c \qquad\qquad\qquad\qquad (9.18)$$

where

D = riser diameter
H = riser height
M_c = casting modulus = V/SA of the casting

These design rules generally are adequate. But for castings with a high modulus, there is a strong possibility of creating a hot spot at the junction between the riser and the casting when a top riser is used. In those instances, the riser dimensions may be increased to ensure that the thermal center is in the riser. Design rules for other shapes, such as tapered risers, have been recently developed by Xia (9).

The feed metal requirement is to ensure that the riser will have sufficient feed metal for the shrinkage volume of the casting. This is usually a binding constraint when the modulus of the casting is low. The theoretical expression for the volume requirements is:

$$V_r = V_c[\alpha/(\beta - \alpha)] \qquad\qquad\qquad (9.19)$$

where

V_r = volume of riser
V_c = volume of casting

β = feed metal capacity of riser (decimal)
α = solidification shrinkage (decimal)

Equation (9.19) assumes that the riser will be fully feeding itself, but that usually does not occur. A better expression assumes that the riser only half feeds itself; that expression is:

$$V_r = V_c[\alpha/(\beta - \alpha/2)] \qquad (9.20)$$

The value of α depends on the particular alloy; the value of the feed metal capacity, β, depends on the riser shape and whether insulating materials are used. Some of the values used are in Tables 9.2 and 9.3.

The feeding path requirements include considerations for directional solidification, minimum temperature gradient, and feeding distance. Directional solidification requires that the modulus continually increase from the edge of the casting to a maximum at the riser. The riser must be the thermal center, and the modulus must continually decrease from the riser to the furthermost section it is to feed. The minimum temperature gradient is somewhat related to a minimum modulus gradient; that is, if the gradient is not steep enough, directional solidification will not occur. The feeding distance expressions are used to help ensure that directional solidification occurs. Some of the basic expressions that were developed for plates and bars for steel castings (6) follow.

Maximum feeding distance—plates $(W \gg T)$:

$$L1 \text{ (mm)} = 72M^{(1/2)} = 140 \qquad (9.21)$$
$$L1 \text{ (in.)} = 14.3M^{(1/2)} - 5.5 \qquad (9.22)$$

Maximum distance between risers:

$$L2 \text{ (in. or mm)} = 4T \qquad (9.23)$$

Maximum feeding distance—bars $(W \approx T)$:

$$L1 \text{ (mm)} = 80M^{(1/2)} - 84 \qquad (9.24)$$
$$L1 \text{ (in.)} = 15.9M^{(1/2)} - 3.3 \qquad (9.25)$$

Table 9.2 Feed Metal Capacity Factors

Riser description	Feed metal capacity (β)
No insulating materials	0.16 (1/6)
Insulating top or sides	0.35
Insulating top and sides	0.65

Table 9.3 Solidification Shrinkage Values

Alloy	Solidification Shrinkage (α)
Steel	0.06
Brass (60-40)	0.045
Copper	0.04
Bronze (88-10-2)	0.065
Aluminum	0.057

Maximum distance between risers:

$$L2 \text{ (in. or mm)} = 2.5T \qquad (9.26)$$

where

$L1$ = maximum feeding distance, in mm or in.
$L2$ = distance between risers, in mm or in.
M = casting modulus of section, in mm or in.
T = casting thickness of section, in mm or in.

For a bar or plate, the modulus of the section can be approximated by the area/perimeter:

$$M = W \times T/2(W + T) \qquad (9.27)$$

Note that if W and T are equal, the expression becomes $T/4$, whereas if W is very large compared to T, the expression approaches $T/2$. If chills are used, the feed distances can be increased by approximately 50 mm, or 2 in. These expressions for feeding distance are for alloys with a narrow freezing range and do not work for alloys with a wide freezing range.

The importance of feeding distance calculations is to determine the number of risers needed to feed the casting. Examples of the type of castings where feeding distance is critical is for long bars or plates or for ring castings. The number of risers also is influenced by the directional solidification paths.

In castings that are chunky, the junction requirement dictates that the modulus of the riser be much larger than that of the casting, up to double that of the casting (6). This is to ensure that the thermal center is in the riser and not at the junction of the riser and the casting. This indicates that side risers should be used rather than top risers in chunky castings to avoid the junction problem.

The pressure requirement and pressure gradient requirements have not been developed into specific design formulas. Three design considerations

(6) that should be followed are: (1) place the feeders so they feed downhill (gravity will assist flow); (2) design feeders so the atmosphere is accessible to assist flow, rather than encourage the formation of a vacuum that will retard flow; and (3) locate gates low, to fill uphill (hot spots may occur near the gate and gates generally do not have feed metal). In instances where the runner is used as a feeder, the gates should be located high so they can feed the casting.

Example Problem 9.2. A steel plate casting, 1 in. × 4 in. × 10 in., is to be fed using a side riser(s) to have a better surface without the riser neck on the center surface. The solidification shrinkage is 6 percent and no insulation aids are used, so the feed metal capacity of the riser is 16 percent, or 0.16. Determine the following:

a.   The number of risers
b.   The dimensions of the riser based upon:
    i. heat transfer requirement
    ii. feed metal requirement

Solution:

a.   The number of risers is indicated by the feeding distance. The casting is a plate and the modulus is:

$$M = W \times T/2(W + T) = 4 \times 1/2(4 + 1) = 0.40 \text{ in.}$$

The maximum feeding distance is:

$$L = 14.3 \times 0.40^{1/2} - 5.5 = 3.5 \text{ in.}$$

    If a center chill is used, this distance can be increased by 2–5.5 in.; thus one riser is required at each end (two risers total are required, one at each end). If the casting length were greater than 11 in., a top riser would be required, to have sufficient feeding. The casting, risers, and chill are illustrated in Fig. 9.7. Each riser will be required to feed one-half of the total casting.

b.   i. The riser dimensions to meet the heat transfer requirement necessitate the modulus calculation. The value of the modulus for each half would be:

$$M_c = \text{volume/cooling surface area}$$
$$= 1 \times 4 \times 5/[2(4 \times 5) + 2(1 \times 5) + 1(1 \times 4)]$$
$$= 0.37 \text{ in.}$$

Note that one of the 1 × 4 sides is connected to the other half and is not a cooling surface.

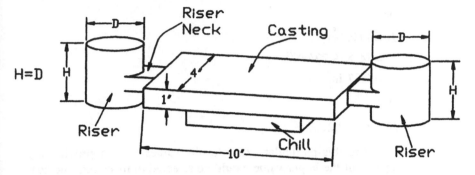

Figure 9.7 Casting, chill, and riser location for Example Problem 9.2.

For a side riser, the riser height equals the riser diameter, which is:

$$D = 6 \times M_c = 6 \times 0.37$$
$$= 2.22 \text{ in.}$$

and

$$H = D = 2.22 \text{ in.}$$

ii. The riser dimensions based on the feed metal requirements necessitate calculation of the riser volume and then determination of the height and diameter based upon the appropriate height and diameter relationship:

$$V_r = V_c[\alpha/(\beta - \alpha/2)]$$

where $V_c = 1 \times 4 \times 5 = 20 \text{ in.}^3$ Thus,

$$V_r = 20[0.06/(0.16 - 0.06/2)]$$
$$= 9.23 \text{ in.}^3.$$

The volume of the riser, as a cylinder, would be:

$$V_r = \pi D^2 H/4$$

But since $D = H$,

$$V_r = \pi D^3/4$$

or

$$D = (4Vr/\pi)^{1/3}$$
$$= (4 \times 9.23/\pi)^{1/3}$$
$$= 2.27 \text{ in.}$$

and

$$H = D = 2.27 \text{ in.}$$

For this particular example, the values were approximately equal, but the larger value would be selected to meet both the heat transfer and volume requirements. The riser dimensions would have the height and diameter equal to 2.27 in., and two risers would be needed.

9.7 YIELD AND ECONOMIC CONSIDERATIONS IN CASTING

The cost of castings is influenced by a large number of factors, such as yield, complexity, shape, and surface requirements. The major cost categories for the foundry are material costs, labor costs, direct foundry expenses, overhead expenses, and environmental costs. The primary material costs for a foundry are the metal costs, the core costs, and the molding sand costs. To illustrate the interaction of the various steps of the casting process upon the costs, an example of how the metal costs are calculated is presented.

The metal cost not only represents the cost of the metal for the casting itself, but also must include costs from scrap caused by core and mold problems, scrap from finishing operation errors, losses in pouring, the recycle of the gates and risers, and environmental costs from dust collection and slag. The key to determination of the metal costs is to develop a consistent calculation scheme (10).

The primary factors needed for estimating the metal cost are:

1. *Casting weight*: weight of the final casting
2. *Yield*: weight of casting plus process returns
3. *Melt loss*: loss due to dross, slag, dust, etc.
4. *Casting scrap rate*: loss due to errors in coremaking, molding, and pouring
5. *Finishing scrap rate*: loss due to errors in grinding and other finishing operations

The values presented are average values, which include the loss values, so these will not correspond to the values poured for a specific casting. For example, the casting is either good or bad, but the average amount of metal poured will include the amount for the good casting plus an amount for the rejected castings. Thus, these calculated amounts are for economic calculations, not for pouring specific individual castings. They are useful for determining the amount to charge the customer for a casting or a batch of castings.

The total weight poured can be calculated by:

$$TWPA = CW \times [1/(1 - CSR)] \times [1/Y]$$
$$\times [1/(1 - ML)] \times [1/(1 - FSR)] \tag{9.28}$$

where

>$TWPA$ = total weight of metal poured on average for casting
>CW = casting weight of finished casting
>CSR = casting scrap rate
>Y = yield
>ML = melt loss
>FSR = finishing scrap rate

For the calculations to be consistent, the total weight poured must also be equal to:

$$TWPA = CW + RW + MLW + CSW + FCSW \tag{9.29}$$

where

>$TWPA$ = total weight of metal poured on average
>CW = casting weight
>RW = return weight of gates and risers
>MLW = melt loss weight
>CSW = casting scrap weight from foundry
>$FCSW$ = finishing and cleaning scrap weight

The calculations relating the return weight, melt loss weight, casting scrap weight, and finishing and cleaning scrap weight are:

$$RW = CW \times [(1 - Y)/Y] \times [1/(1 - CSR)] \times [1/(1 - FSR)] \tag{9.30}$$

$$MLW = CW \times [ML/(1 - ML)] \times [1/Y]$$
$$\times [1/(1 - CSR)] \times [1/(1 - FSR)] \tag{9.31}$$

$$CSW = CW \times [CSR/(1 - CSR)] \times [1/(1 - FSR)] \tag{9.32}$$

$$FCSW = CW \times [FSR/(F - FSR)] \tag{9.33}$$

Example Problem 9.3. An engineer has designed a product as a casting that has a total volume of 84 in.^{3}. The metal being used is cast iron, which has a density of 0.25 lb/in^3. The casting engineer has determined that the foundry has induction melting and green sand molding and estimates the various factors as:

Factor description	Amount	(Percent)
Yield	0.65	(65%)
Melt loss	0.03	(3%)
Casting scrap	0.10	(10%)
Finishing scrap rate	0.04	(4%)

If the hot metal costs are \$4.00 per pound, the environmental costs are \$15.00 per pound, and the credit for scrap and returns is \$1.00 per pound, what should be the total metal cost for the casting?

Solution. The solution procedure is first to determine the casting weight and then to determine the various components of the total weight poured. The casting weight would be:

$$CW = 84 \text{ in.}^3 \times 0.25 \text{ lb/in.}^3 = 21 \text{ lb}$$

The total weight poured can be determined from Eq. (9.27):

$$TWPA = 21 \times [1/(1 - 0.10)] \times [1/0.65] \times [1/(1 - 0.03)]$$
$$\times [1/(1 - 0.04)] = 38.55 \text{ lb}$$

The components of the total weight poured are:

CW = casting weight = 21.00 lb

RW = return weight

$$= 21 \times [(1 - 0.65)/0.65] \times [1/(1 - 0.10)]$$
$$\times [1/(1 - 0.04)] = 13.09 \text{ lb}$$

MLW = melt loss weight

$$= 21 \times [0.03/(1 - 0.03)] \times [1/0.65] \times [1/(1 - 0.10)]$$
$$\times [1/(1 - 0.04)] = 1.16 \text{ lb}$$

CSW = casting scrap weight

$$= 21 \times [0.10/(1 - 0.10)] \times [1/(1 - 0.04)] = 2.43 \text{ lb}$$

$FCSW$ = finishing and cleaning scrap weight

$$= 21 \times [0.04/(1 - 0.04)] = 0.88 \text{ lb}$$

The total weight of the poured casting is:

$$21.00 + 13.09 + 1.16 + 2.43 + 0.88 = 38.56 \text{ lb}$$

The difference in the two values of 38.55 and 38.56 is due to rounding in the determination of the four components that were added to determine the casting weight. The costs of the metal would be:

$$\text{Hot metal costs} = 38.56 \text{ lb} \times \$4.00 = \$154.24$$
$$\text{Environmental costs (melt losses)} = 1.16 \text{ lb} \times \$15.00 = 17.40$$
$$\text{Returns credit (gates and risers)} = 13.09 \text{ lb} \times \$1.00 = -13.09$$
$$\text{Casting scrap credit (castings)} = 2.43 \text{ lb} \times \$1.00 = -2.43$$
$$\text{Finishing scrap credit (castings)} = 0.88 \text{ lb} \times \$1.00 = \underline{-0.88}$$
$$\text{Total metal cost} = \$155.24$$

This example problem indicates the large difference between the total weight of metal needed to make the casting and the total weight of the casting. The calculation procedure permits an economic evaluation of scrap rates and the importance of process control.

9.8 SUMMARY OF THE CASTING PROCESS

Casting is the easiest way to go from raw material to finished product. It has several advantages over other production methods, mainly the reduction or elimination of machining operations. There are many different casting processes, which can be classified according to mold type, pattern type, and production type. In addition to the design of the casting shape, the design of the gating and riser/feeding systems are important factors. The pattern design involves not only the design of the part, but also consideration of the shrinkage that occurs during the solidification and cooling of the metal. In addition to the functional design of the part, the thermodynamic effects during solidification and the fluid effects of the metal during mold filling make the casting design a complex and challenging process.

The economics of the casting process is greatly affected by process yields and scrap rates. One major advantage of the casting process is that the scrap metal and returns can be recycled immediately. Also, the metal-casting processes tend to use large quantities of scrap materials and thus to contribute positively to many environmental issues.

9.9 EVALUATIVE QUESTIONS

1. How are casting processes classified?

2. What is meant by *green sand casting*?

3. What is an approximate value for the process yield of a casting? Why is this value not 100 percent?

4. What are the seven major types of defects that the foundry has with castings?

5. What is the major function of the gating system?

6. Explain the characteristics of a pressurized gating system, making reference to the cross-sectional areas and the flow velocities.

7. Give a description of one of the casting processes in Table 9.1, and discuss the major advantages and limitations of that process.

8. Using the iron–iron carbide phase diagram (Fig. 2.6) for a cast iron of 3.0°C cast from 1400 °C, estimate the amount of superheat and the freezing range for the alloy.

9. Using the general phase diagram (Fig. 2.3) for a 25 percent B alloy cast at 3200°C, estimate the amount of superheat and the freezing range of the alloy.

10. A steel casting 1 in. × 4 in. × 12 in. is poured in 10 sec, the effective sprue height is 4 in., and the gating ratio is 1:2:3. The density of steel is 7.8 g/cm³ (0.28 lb/in.³) and the casting yield is 60 percent. The tapered cylindrical sprue is connected to two square runners, and each runner is connected to two gates with width twice the height.
   a. What is the amount of metal poured (in lb and in.³)?
   b. What is the pouring rate (in lb/sec and in.³/sec)?
   c. What is the choke area (in in.²), and where is the choke?
   d. What is the choke velocity (in in./sec)?
   e. What are the dimensions of the sprue base, each runner, and each ingate?
   (a. 22.4 and 80; b. 2.24 and 8; c. 0.144; d. 55.6; e. 0.43, 0.38 × 0.38, 0.23 × 0.46)

11. Using the data of Question 10, and assuming the feed metal fraction for the sand riser is 16 percent and the shrinkage is 6%, determine the riser dimensions for a side riser:
   a. Based upon the solidification time
   b. Based upon the feed metal requirements
   c. Determine the feeding distance of the riser, and determine whether it is sufficient.
   (a. 2.25 in.; b. 3.04 in.; c. 3.5 in., no)

12. A steel plate casting, 1 in. × 4 in. × 12 in., is to be poured in 5 sec. The runner is 4 in. below the top of the cope, and the gating ratio is 1:3:3 with one down sprue, two runners, and four ingates (gates). Assume the sprue is a tapered cylinder, the runners are square, and the gates are rectangular with the width twice the height. The density of steel is 7.8 g/cm³, the dynamic viscosity is 6 centipoise, and the estimated yield is 50

percent. The solidification shrinkage is 5 percent, and the riser feed metal
fraction is 16 percent. Use a top riser for feeding.

   a. Make a sketch of the casting and gating system, and calculate the
      dimensions of the sprue base, runner, and ingates.
   b. Recommend the size of the riser, and check the thermal adequacy,
      the volume of feed metal, and the feeding distance requirements.
   c. What is the maximum Reynolds number?
      (answers)

$V_{ch} = 55.6$ in./sec	D(thermal) $= 2.25$ in.
$D_s = 0.66$ in.	D(feed) $= 3.56$ in. (approx. 4)
$W(r) = H(r) = 0.72$ in.	FD $= 10.08$ in. (approx.)
$W(g) = 0.72, H(g) = 0.36$ in.	$R\#$ $= 31,000$

13. Rework Question 12 with a gating ratio of $3:2:1$, a yield of 60 percent,
and an effective sprue height of 2 in.
(answers)

$$D_s = 1.25 \text{ in.}$$
$$W(r) = H(r) = 0.64 \text{ in.}$$
$$W(g) = 0.45 \text{ in.}, H(g) = 0.225 \text{ in.}$$
$$R\# = 10,000$$

14. A design engineer has designed a product as a casting that has a total
volume of 100 in.^{3}. The metal used is aluminum, which has a density of
0.10 lb/in.^{3}. The metal is melted in a gas furnace, green sand molding is
used, and the estimates of the various yield and scrap factors are:

Yield	45%
Melt loss	5%
Casting scrap	10%
Finishing scrap rate	2%

If the hot metal costs are $3,00/lb, the environmental costs are $10.00/
lb, and the credit for scrap and returns are $0.50/lb, what should be the total
metal cost for the casting?
(casting weight 10 lb; pouring weight 26.52 lb; runner weight 13.86 lb;
casting scrap 1.13 lb; finishing scrap 0.20 lb; melt loss 1.33 lb; Cost =
$85.26)

15. A circular disc is to be cast in steel (Fig. 9.8). The steel has a solidification shrinkage of 6 percent, feed metal capacity of 15 percent, density
of 0.28 lb/in.^{3}, and an estimated yield of 60 percent. The gating ratio is
$1:3:5$, and a single tapered sprue is used with two square runners and four

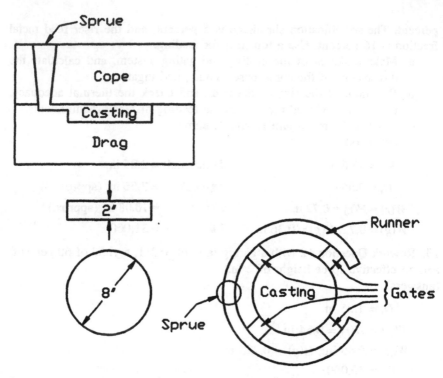

Figure 9.8 Casting and gating system for Evaluative Question 15.

gates with a height:width ratio of 1:3. The circular disc is 8 in. in diameter and 2 in. thick. The effective sprue height is 6 in., and the desired pouring time is 15 sec. A top cylindrical riser is to be used and is located at the center of the disc.

   a. What is the modulus of the casting, in inches?

   b. Estimate the diameter of the riser based upon solidification time requirements.

   c. Estimate the diameter of the riser based upon feed metal requirements.

   d. Determine the pouring rate (in.^{3}/sec) for the casting.

   e. Determine the diameter of the sprue base.

REFERENCES

1. Simpson, Bruce Liston. History of the Metalcasting Industry, 2nd ed., 1969, American Foundrymen's Society, pp. 6–9.

2. U.S. Department of Commerce, International Trade Administration, Office of Industrial Resource Administration, Strategic Analysis Division. Investment Castings: A National Security Assessment, December 1987, pp. 108–109.
3. 1995 Casting Design and Application Reference Handbook, Penton Publication, pp. 24–31, 50.
4. Todd, Robert H., Allen, Dell K., and Alting, Leo. Manufacturing Processes Reference Guide, 1994, Industrial Press, pp. 230–250.
5. Rowley, Mervin T., ed. and Trans. International Atlas of Casting Defects, 8th ed., 1974, International Committee of Foundry Technical Associations, American Foundrymen's Society, p. 7.
6. Campbell, John. Castings, 1991, Butterworth-Heinemann Ltd., London, pp. 22, 179–191.
7. Kondic, V. Metallurgical Principles of Founding, 1968, American Elsevier Publishing Company, New York.
8. Creese, R. C. "Optimal Riser Design by Geometric Programming," AFS Cast Metals Research Journal, 1971, vol. 7, no. 4, pp. 182–185.
9. Creese, R. C., and Xia, Y. "Tapered Riser Design Optimization," AFS Transactions, Vol. 99, 1991, pp. 717–727.
10. Creese, R. C., Adithan, M., and Pabla, B. S. Estimating and Costing for the Metal Manufacturing Industries, 1992, Marcel Dekker, New York, pp. 177–194.
11. Alting, Leo. Manufacturing Engineering Processes, 2nd ed., 1994, Marcel Dekker, New York, pp. 13, 301–341.

BIBLIOGRAPHY

Beeley, P. R. Foundry Technology, 1972, Butterworth Scientific, London.
Casting, Volume 15: Metals Handbook, 9th ed., 1988, ASM International, Metals Park, Ohio.
Flinn, R. A. Fundamentals of Metal Casting, 1963, Addison-Wesley, Reading, MA.
Heine, R. W., Loper, C. R., and Rosenthal, P. C. Principles of Metal Casting, 2nd ed., 1967, McGraw-Hill, New York.
Webster, P. D., Ed. Fundamentals of Foundry Technology, 1980, Portcullis Press Ltd., Redhill, England.

INTERNET SOURCES

http://www.implog.com/foondry/
http://web.staffs.ac.uk/sands/engs/des/aids/process/welcome.htm
http://www.cemr.wvu.edu/~imse304/
http://amc.scra.org/

10
Plastic Parts Manufacturing

10.1 INTRODUCTION

Plastics manufacturing methods are similar to those for metal casting, in that a liquid or semiliquid (plasticate) is injected into a mold to form a shape. In particular, plastic injection molding resembles the die casting process, for both processes use high pressures and produce parts of the same general size and shape. Plastic injection molding resembles the powder metallurgy process, in that the initial starting material is often small particles or granules heated to form the liquid or semiliquid.

Plastics are one of the most versatile family of materials available for making products. Plastics applications have grown tremendously during the last 30 years. The major reasons for the increased use of plastics include light weight, chemical resistance, electrical and thermal insulation properties, corrosion resistance, low cost, high production quantities, appearance, and ease of processing. Plastics are rapidly replacing metals in automotive, aerospace, and domestic applications. Compared with metals, plastics have lower weight (one-eighth that of steel), lower energy requirements for processing, and higher resistance to chemicals and the environment, and their strength can be increased by making composites with fibers.

Plastics, or *polymers*, are materials made from a number of smaller molecules called *monomers*. Monomers are linked together to form long chains of polymers with useful properties. For example, ethylene (C_2H_4) is a monomer that is polymerized with the help of catalysts, heat, and pressure to create high-molecular-weight chainlike molecules of polyethylene.

This chapter was written by Dr. Sheikh Burhanuddin; minor modifications were made during review and editing.

```
  H  H                        H  H  H  H  H
  |  |                        |  |  |  |  |
 -C==C-                      -C==C==C==C==C-
  |  |                        |  |  |  |  |
  H  H                        H  H  H  H  H
```

ETHYLENE MONOMER   POLYETHYLENE POLYMER

There is a wide range of natural and synthetic polymers. Natural polymers exist in plants and animals, and include proteins, cellulose, and natural rubber. Synthetic polymers are derived mainly from mineral oil, natural gas, and coal. Some of the more common synthetic polymers include nylon, polyethylene, phenolics, and epoxies.

10.2 CLASSIFICATION OF POLYMERS

The two major classification methods for polymers are: (1) characteristics at elevated temperatures, and (2) chemical families having the same monomer.

On the basis of behavior at elevated temperatures, polymers are classified as thermosets and thermoplastics. *Thermoset plastics* cannot change shape after being cured or polymerized. *Thermoplastics* are solid at room temperature but soften and eventually melt as temperature is increased. Thermoplastics can be melted and resolidified many times, like metals. Thermosets are harder, stiffer, and chemically more inert than thermoplastics. Common examples of thermosets are: phenolics, epoxies, alkyds, and melamine. Common examples of thermoplastics are: polyethylene, polyvinylchloride, nylon, and polystyrene. Plastics are often mixed with fillers, pigments, and stabilizers to aid in processing and for enhancing their properties.

Classification of polymers into chemical families having the same monomer helps in grouping most polymers into a limited number of families instead of considering thousands of polymers individually. For example, in the chemical family based on ethylene monomer, different polymers, such as polypropylene and polystyrene, can be created by substituting some of the hydrogen atoms on the monomer with functional groups such as CH_3 (for polypropylene), OH (for polyvinyl alcohol), and C_6H_5 (for polystyrene).

10.3 PLASTIC PARTS MANUFACTURING PROCESSES

The major processes for making plastic parts are presented in a list and then described in more detail. The six major processes are:

Injection molding
Compression molding
Blow molding
Extrusion
Thermoforming
Plastic reinforcing and composites

10.3.1 Injection Molding

Injection molding is by far the most common process for making plastic parts. The major reasons for its popularity are shape controllability, accurate dimensions, and high production rate. The process is used primarily for thermoplastics but it may be applied to thermosets. Equipment and mold costs are high. But large volumes and high production rates make this process very economical.

An injection molding machine makes formed parts from polymeric materials. The material is fed through a hopper into a barrel that houses a screw and heaters. The rotating screw advances the material toward its tip into a pressure chamber. The heaters are used to heat and plasticate the material as it advances through the barrel. When sufficient material accumulates in the pressure chamber for one shot, the screw automatically advances and injects the material into a clamped mold, where it solidifies in the shape of the cavity. A mold usually consists of a sprue, runner, and gating system along with one or more cavities. Upon solidification of the material in the mold, the mold is opened and parts are removed by ejection pins. The mold is clamped again and the cycle is repeated. A typical injection molding machine is shown in Fig. 10.1. Three types of simple mold construction are shown in Fig. 10.2. Various configurations of these three systems are utilized in mold making.

10.3.2 Compression Molding

Compression molding is used mostly for making parts from thermosetting plastics. Thermosetting plastic in the form of powder or a preformed tablet is placed in a heated cavity and the mold closed under pressure. Molds may be heated by steam or electric heating coils. A binding agent may be added to the material to form a reinforced plastic part. Transfer molding is similar to compression molding except that the material is preheated into a liquid form before being forced into the mold. Transfer molding is suitable for making intricate shapes and where metal inserts are molded into plastic. The major benefits of compression molding are a very low tendency toward distortion and warpage and a high degree of part density.

Figure 10.1 A typical injection molding machine (Courtesy of Cincinnati Milacron, Batavia, OH.)

10.3.3 Blow Molding

Blow molding is used mostly for making hollow shapes such as containers, bottles, automobile fuel tanks, and refrigerator liners. This process involves placing a heat-softened plastic extruded tube or parison in a two-piece mold, closing its end, and inflating it with compressed air so that it takes the shape of the mold. In blow molding, the problems of weld and flow lines and mold erosion are reduced because the material within the mold stretches rather than flowing. Blow-molded parts with undercuts can be more easily removed than injection-molded parts. Tolerances on wall thickness are not tight. Blow molds usually cost less than injection molds. Figure 10.3 shows an automobile fuel tank produced by blow molding. Typical blow-molded automobile parts are illustrated in Figure 10.4

10.3.4 Extrusion

Extrusion is used for making thermoplastic parts of continuously uniform cross sections. The process is similar to injection molding except that instead of injecting the material into a mold, the material is extruded through a die to obtain a desired form. Extrusion is used for making structural shapes

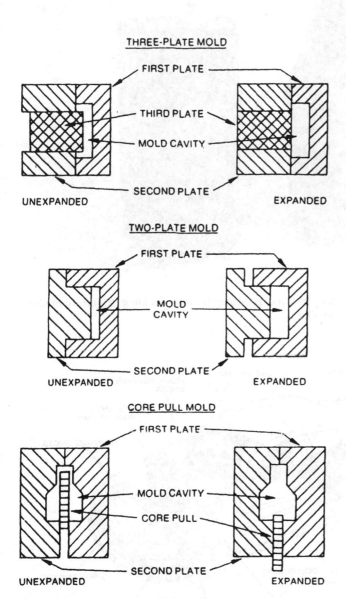

Figure 10.2 Three types of simple mold construction. (Reprinted from Ref. 3, p. 130, by courtesy of Marcel Dekker, Inc.)

Figure 10.3   Automotive fuel tank produced by blow molding. (Courtesy of Cincinnati Milacron, Batavia, OH.)

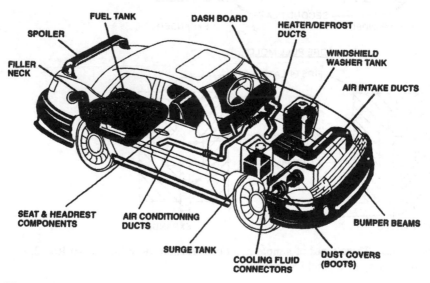

Figure 10.4   Automotive blow-molded parts. (Courtesy of Cincinnati Milacron, Batavia, OH.)

(channels, bars, angles, etc.), pipes, sheets, film, wire coverings, cable sheathing, and fibers. An extruding machine is shown in Fig. 10.5.

One variation of the extrusion process is *coextrusion*, which involves extruding layers of two or more different polymers simultaneously. Each polymer type contributes some desired property. Coextrusion examples include beverage cups, containers, refrigerator liners, and foam-core solid sheath wires. Figure 10.6 shows a coextruded beverage cup.

10.3.5 Thermoforming

Thermoforming is a thermoplastic sheet-forming technique used for making cuplike shapes. In this process, a premanufactured thermoplastic sheet is clamped, heated, and shaped over or into a mold. Trimming is often required after forming. Forming is accomplished using vacuum, pressure, matched (male) molds, and their combinations. This process is fast and can be easily automated for long production runs. It is a very cost-effective technique because of fast cycle times and low mold costs. Major uses of the thermo-

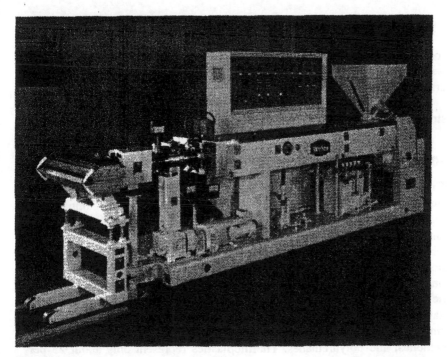

Figure 10.5 Sheet and film extrusion machine. (Courtesy of Wilex Corp., Blue Bell, PA.)

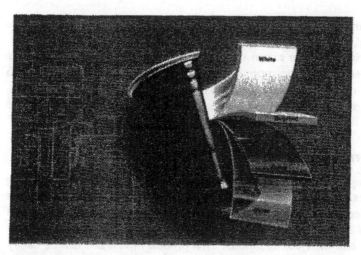

Figure 10.6 Thermoformed cup made from a four-layer coextruded sheet. (Courtesy of Wilex Corp., Blue Bell, PA.)

forming process include packaging, food trays, tumblers, contoured plastic windshields, bus and aircraft seat backing, refrigerator and freezer door liners, and luggage.

Excessively deep-drawn parts and small-radius curves tend to overstretch the sheet. Holes in thermoforming products should also be avoided.

10.3.6 Plastic Reinforcing and Composites

The mechanical properties of plastics (stiffness, toughness, tensile and compressive strength, and resistance to cracking, creep, fatigue, impact, and abrasion) can be increased substantially and resistance to mold shrinkage can be reduced by adding reinforcing fibers such as glass, cotton, paper, carbon, nylon, and kevlar. Fibers in reinforced plastics are generally in short pieces and are randomly distributed. Composites, on the other hand, have long, unbroken strains of fiber. Even though reinforced plastics do not have the load-carrying capability of composites, they make excellent structural materials for a large number of products. Most of the reinforced plastics and composites contain thermosetting-type resins such as polyester, epoxies, phenolics, and polyurethanes. Thermoplastics represent only about 25 percent of the total reinforced-plastics and composites market. Major techniques used for reinforcing plastics include hand layup, spray layup, and compression and injection molding. The techniques for making composites include pultrusion, filament winding, and laminating. Reinforced plastics and composites are used for making a variety of products, such as boat hulls, bathtubs, aircraft and automobile parts, printed circuit boards, fishing rods, ladders, structural shapes, and pipe.

10.4 PLASTIC PROCESSES AND PRODUCTS FROM VARIOUS POLYMERS

Table 10.1 shows a listing of manufacturing processes for and typical uses of various types of polymers. The polymers can be shaped by more than one process; thus, one must specify both the polymer and the process. The typical uses illustrate the wide variety of applications of the various polymers. The products illustrated also emphasize the primary advantages of polymer products: light weight, insulating properties for electrical products, thin walls for containers and packaging, and high production capabilities. The wide variety of products have generally been developed for low-load applications, but reinforced-fiber polymers are being used in structural applications such as reinforcing bars and structural box and beam shapes.

10.5 DESIGN AND MANUFACTURING CONSIDERATIONS FOR PLASTIC PARTS

Complex molds can be made to produce complex parts but at higher costs and longer design times. The designer should try to keep the part design simple because it translates into good molds. A two-phase design approach is recommended. In phase 1, the designer determines the basic part shape and various features. In the second phase, the designer systematically evaluates each part shape feature for ease of manufacturing at lower cost. A set of established design guidelines can be very useful in accomplishing this task. Most part features consist of combinations of three basic shape elements: nominal wall, projections off walls (ribs, etc.), and depressions into walls (1).

Some recommended design guidelines to improve plastics part manufacturability, reduce potential manufacturing problems, and molding cost follow.

Wall design: The basic function of walls is to bear load and support other part shape features. Wall thickness should be kept as uniform as possible. Thick and thin wall sections cool at different rates. Thick sections take longer to cool and cause voids, sinks, warpage, and stress buildup. Gradual transition should be allowed when thin and thick sections must be mixed.

Table 10.1 Manufacturing Processes for and Typical Uses of Various Polymers

Type of polymer	Manufacturing processes	Typical uses
Acrylonitrile-butadiene-styrene (ABS)	Injection and blow molding, extrusion, thermoforming	Telephones, luggage, pipe fittings, appliances
Acrylic	Injection molding, extrusion into formed shapes and sheets	Outdoor signs, instrument panels, auto parts, packaging
Alkyd	Compression and injection molding	Ignition parts, switches, circuit breakers, appliance parts
Epoxy	Compression and injection molding	Bobbins for windings, electronics components, epoxy adhesives for joining various materials
Nylon	Injection molding, extrusion	Bristles for paintbrushes and toothbrushes, gears, cams, bearings, radiator fans, timing sprockets, packaging
Phenolic	Injection and transfer molding	Electrical parts (switch gears, relays, connectors, etc.), auto parts (brake systems, transmission parts, valves, etc.), handles and knobs for utensils and small appliances
Polycarbonate	Injection and blow molding, extrusion, thermoforming	Auto parts (lamp lenses and housings, electrical parts, body parts, etc.) milk bottles, mugs, pitchers, circuit boards, lighting applications, safety shields

Polyethylene	Injection, blow, and roto molding, extrusion, thermoforming	Low-density polyethylene (LDPE) for packaging film, wire and cable insulation, blow-molded containers; high-density polyethylene (HDPE) for large roto-molded parts, containers for milk and chemicals
Polypropylene	Injection and blow molding	Storage battery cases, carpeting, packaging, housewares, auto parts
Polystyrene	Injection and blow molding, extrusion, thermoforming	Containers for drugs, food, and produce, disposable dishes, refrigerator liners, furniture components, housewares
Polyvinyl chloride (PVC)	Extrusion, blow molding, thermoforming	Pipe, building materials (house siding, gutters, electrical conduits, floor tile), water supply and sanitary systems, furniture

Ribbing and coring on thin walls should be used to replace thick walls. Beck (2) presents recommended wall thicknesses for thermoplastic and thermosetting plastics.

Ribs: Ribs are recommended for strengthening walls, for improving stiffness, and for supporting other components. Figure 10.7 shows proportional dimensions of ribs, which result in minimization of voids, sink marks, and stresses and which improve molding. It is better to use a series of thin ribs rather than one heavy rib.

Radii: All intersections must have rounded surfaces. Sharp corners concentrate stress, inhibit material flow, and cause brittle failure of material. Sharp inside corners cause material failure under impact due to notch sensitivity. The inside corner radius should be at least one-fourth of wall thickness, whereas the outside corner radius should be at least 1.25 times the wall thickness. A sharp corner is usually easier to produce in a mold, but it is highly undesirable. Fillets should be used at the base of ribs and bosses to strengthen them.

Taper and draft angles: Taper or draft facilitates part removal from the mold. The taper should be given on all vertical walls including the core. Generally a 1° taper is adequate for small parts. Large and deep parts require larger taper. As a general rule the taper should be as large as the functional requirements of the part will allow.

Inserts: Inserts in plastic parts should be avoided as much as possible. When an insert must be used, the designer should make sure that inserts do not present any sharp corners to the plastic. The amount of plastic around an insert should be thick enough not to crack.

Undercuts: Undercuts should be avoided whenever possible. Undercuts can be realized by providing sliding components or split cavity cam actions in the mold, but these provisions increase the cost of the mold. Parts

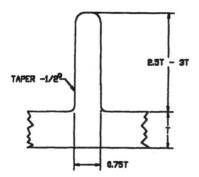

Figure 10.7   Design guidelines for ribs including taper, length, and thickness.

with shallow undercuts and coarse threads con be stripped conventionally, with proper precautions.

10.6 ESTIMATING COSTS FOR AN INJECTION-MOLDED PART

A rough estimate of the cost of an injection-molded part is often required before doing detailed design work. An example initially developed by Wendle (3) has been modified to have four major cost components:

1. Material cost
2. Setup cost
3. Molding cost
4. Labor cost

Each of the components is presented in more detail for making rough or conceptual cost estimates.

1. *Material cost*: The first step in determining the material cost is the estimation of the cubic inches of material the part will use. The estimated part volume in cubic inches is multiplied by 0.036 and 1.5 to determine the weight of the material used. The specific weight of water is 0.036 lb/in.^{3}, or 0.001 kg/cm^3 if the metric system is used, and the average specific gravity for plastics may be taken as 1.5. An 80 percent yield factor (i.e., 20 percent sprue, runners, and gates) is commonly used in industry, and a 10 percent scrap factor will be used for illustrative purposes. The cost of material is obtained by multiplying the weight of the part, in pounds (kilograms) by the cost per pound (per kilogram) of the material. The cost of material per pound may be obtained from material suppliers or publications (4). In the absence of material cost data, one may use $1.50/lb ($.70/kg) for commodity resins (styrene, polypropylene, polyethylene), and $3.00/lb ($1.40/kg) for common engineering resins (ABS, polycarbonate).

2. *Setup cost*: An average setup charge of $200 per 1000 parts, or $0.20 per part, is common in industry. This includes the labor costs for the setup.

3. *Molding cost*: The molding cost depends on the machine size and estimated cycle time. Table 10.2 gives machine size and hourly machine cost for a given part weight in ounces and in grams. For a multiple-cavity mold, substitute the total weight of material in all cavities for the part weight in Table 10.2. Cycle time is determined using Table 10.3. For a given wall thickness of the part, Table 10.3 gives the cycle time in seconds. The cycle time obtained from Table 10.3 is multiplied by the average hourly machine rate obtained from Table 10.2 to get the molding cost. The molding cost is added to the material, labor, and setup costs to determine the part cost.

Table 10.2 Machine Rate as a Function of Part Weight and Machine Size

Part weight		Machine size (tons)	Average machine rate ($/hr)
Ounces	Grams		
<5	<140	50–75	25
5–10	140–285	100	35
10–20	285–570	200	50
20–40	570–1140	300	70
40–60	1140–1700	400	80
60–90	1700–2550	500	90
90–150	2550–4250	700	100
150–250	4250–7100	1000	120

Reprinted from Ref. 3, p. 90, by courtesy of Marcel Dekker, Inc.

4. *Labor cost*: The labor cost is determined from the operator rate and the unit time to produce a part.

10.6.1 Cost Estimating Example

The following example problem follows a general approach presented by Wendle (3). The data is given for a lot of 2000 parts. The total part volume is 36 in.^{3}, the anticipated yield is 80 percent, and the scrap rate is 10 percent. The mold has four cavities. The part material is polycarbonate, which is available at $3.00/lb. The part has a maximum wall thickness of 0.1875 in.

Table 10.3 Cycle Time as a Function of Wall Thickness

Thickest wall section		Cycle time (sec)	Shots/hr
in.	mm		
0.040	1.00	13	277
0.050	1.25	15	240
0.100	2.50	25	144
0.120	3.00	32	112
0.150	3.75	38	95
0.175	4.50	48	75
0.200	5.00	60	60

Reprinted from Ref. 3, p. 91, by courtesy of Marcel Dekker, Inc.

The density of part material is 0.045 lb/in.^{3}. Estimated setup time is 2 hr at $200/hr, and the operator rate is $24/hr.

part cost = material cost + setup cost + molding cost + labor cost

Material cost:

part material weight = 36 in.^{3} × 0.045 lb/in.^{3} = 1.62 lb = 26 oz

part weight adjusted for yield and scrap = 1.62 lb
   × 1/0.8 × 1/(1 − 0.1) = 2.25 lb

material cost (including yield and scrap adjustments)
   = 2.25 lb × $3.00/lb = $6.75/part

Setup cost:

unit setup cost = $200/hr × 2 hr/2000 parts = $0.20/part

Molding cost:

material weight in four cavities = 2.25 lb × 4 = 9.0 lb = 144 oz

From Table 10.2, the recommended machine size is 700 tons and the average hourly machine rate is $100/hr. From Table 10.3 the cycle time is 60 sec so the molding cost is:

$$\text{molding cost} = \$100/\text{hr} \times \frac{60 \text{ sec}}{3600 \text{ sec/hr}}$$

$$= \$1.67/\text{molding cycle}$$

$$\text{molding cost per part} = \frac{\$1.67 \text{ cycle}}{4 \text{ parts/cycle}} = \$0.42/\text{part}$$

Labor cost:

labor cost = $24/hr × 60 sec/cycle × 1 cycle/4 parts
   × 1hr/3600 sec
   = $ 0.10/part

total unit cost = 6.75 + 0.20 + 0.42 + 0.10 = $7.47
production time = 60 sec/cycle × 1 cycle/4 parts × 1/(1 - 0.1 scrap)
(2000 part lot)   × 2000 parts × 1 hr/3600 sec + 2 hr (setup)
   = 9.26 hr + 2.000 hr = 11.26 hr

These costs and production times are only conceptual estimates, that is, within a range of −30 to + 50 percent of the actual cost.

10.7   TROUBLESHOOTING PART DEFECTS

A brief list of typical molding problems (defects) and potential causes of the defects in plastics manufacture is presented. The potential causes frequently are key process variables that must be controlled to prevent the defects from occurring. Some of these variables are process variables, such as mold temperature, material temperature, injection pressure, ram speed, and clamping pressure. Other causes are design problems, such as too small runners and gates, abrupt section variations, and poor rib design.

> *Problem*: short shots or surface wrinkles
> *Potential causes*: low injection pressure; low mold temperature; inadequate mold venting; small runner or gates
>
> *Problem*: sunken areas opposite a perpendicular rib
> *Potential causes*: low injection pressure; high material temperature; clogged vents; low ram speed; inappropriate rib dimensions
>
> *Problem*: surface streaking
> *Potential causes*: low mold temperature; material not dry enough; high injection rate; short corners around gates
>
> *Problem*: warping of parts
> *Potential causes*: abrupt section variations; high mold temperature; high material temperature; low ram pressure time
>
> *Problem*: excessive flash around parting lines
> *Potential causes*: high injection speed; low clamping pressure; high injection pressure; high material temperature
>
> *Problem*: dimensions too large
> *Potential causes*: high injection pressure; low mold temperature
>
> *Problem*: dimensions too small
> *Potential causes*: high mold temperature; high material temperature; low injection pressure; short ram pressure time
>
> *Problem*: shot-to-shot dimension variation
> *Potential causes*: mold temperature not constant; material temperature not constant; injection pressure variation

10.8   SUMMARY

The use of plastics in industrial and domestic applications has grown rapidly during the last two decades. This trend is attributable to the availability of a large variety of plastics with desirable properties.

Plastics are classified into two major classes: thermosets and thermoplastics. Thermoset plastics can be softened with heat and formed into a shape only once, whereas thermoplastics can be softened with heat and formed into a shape repeatedly. Major manufacturing processes available for making plastics parts include injection molding, extrusion, blow molding, thermoforming, compression molding, and plastics reinforcing and composites. The design of a plastic part must be compatible with the manufacturing process used for making the part. Guidelines for design and manufacturing of plastics parts help reduce potential problems in the manufacturing and use of parts.

10.9 EVALUATIVE QUESTIONS

1. What is a monomer?

2. What is the difference between a natural and a synthetic polymer? Give an examples of each type of polymer.

3. How are plastics classified based on their temperature characteristics?

4. What are the main characteristics of a thermoplastic polymer?

5. What are the main characteristics of a thermosetting polymer?

6. How can different polymers be synthesized from same monomer?

7. a. What are the six manufacturing processes for producing plastic parts?
   b. Which processes use thermoplastic materials and which processes use thermosetting materials?

8. A 25-in.^{3} plastic part is to be made by injection molding. The cost of material is \$2.00/lb, and its specific gravity is 1.4. The mold has four cavities. The part has a maximum wall thickness of 0.15 in. Determine the unit cost if the setup time is 1 hr, the setup cost is \$200/hr, the labor rate is \$20/hr, the material scrap rate is 20 percent, and the lot is 1000 parts. (each 0.036 lb/in.^{3})

9. A 450-cm^3 plastic part is made by injection molding. The part material has a density of 1.2 g/cm^3, and the cost is \$5.00/kg. The mold has three cavities, the setup time is 1.5 hr, the setup rate is \$200/hr, the wall thickness is 4.50 mm, the material scrap rate is 20 percent, and the labor rate is \$36/hr. If the lot is 3000 parts, what is the lot cost and the unit cost?

REFERENCES

1. Beall, G. L. Plastic Part Design for Economical Injection Molding, Borg-Warner Chemicals, Parkersburg, WV, 1983, p. 20.
2. Beck, R. D. Plastic Product Design, 2nd ed., Van Nostrand Reinhold, New York, 1980, p. 126.
3. Wendle, B. C. Developing Plastics Products, Marcel Dekker, New York, 1991, pp. 89–91, 130.
4. Kreisher, K. R., ed., Modern Plastics, 25:1, McGraw-Hill, New York, 1995, p. 63.

INTERNET SOURCES

http://xenoy.mae.cornell.edu/index.html
http://www.matrixplastics.com/quickdirt.htm
http://www.bpf.co.uk/publicat.htm

11
Powder Processing

11.1 INTRODUCTION

Powder processing—in particular, powder metallurgy—is a near-net-shape–
forming process. It is a primary shape-forming process, but unlike casting,
forming, and machining, it is not classified as one of the major shape-forming processes. Sometimes it is classified as a casting process, since some
powders, such as ceramics, are in the form of slurries and are poured into
a mold. On the other hand, the powder is pressed into a die to form the part
shape, and this is more similar to the bulk deformation processes in forming.
In reality, it is a separate major shape-generating process and is included in
this book in a separate chapter between casting and forming. It has distinct
advantages over casting and forming, such as higher yields and compositions
that cannot be obtained otherwise. It is used extensively for the forming of
small, thin, complex parts that are difficult to form with the other processes.
The main disadvantages are the high costs of the materials, since they must
be in powder form, and the inability to form large parts because of press
size limitations. The emphasis of this section will be on powder metallurgy
rather than on materials such as glass and ceramics.

Although powder metallurgy (PM) is seen as a new, high-technol-
ogy process, it has been around for centuries. Prior to the "discovery" of
America, powder metallurgy was used by the Incas for making jewelry; they
used platinum in jewelry, and there were no containers to hold liquid platinum until the 1800s. Platinum melts at 1770°C (3218°F). Thus, the Inca
Indians were able to produce jewelry via powder metallurgy long before the
melting of platinum in crucibles was possible. This is one of the early "high-
technology" innovations of the "New World"; however, most of the technology was lost as the "Old World" colonized the "New World."

Modern powder metallurgy was started about 1910 with the production
of ductile tungsten for light bulb filaments. The first commercial powder
metallurgy parts were porous bronze bushings produced in the 1920s; the

first ferrous PM parts were oil pump gears manufactured in the 1930s (1). Stainless steel, aluminum, high-speed steels, and cemented carbide parts are also commonly produced via powder metallurgy, as are parts from low-carbon steels and copper-base alloys. For materials with high melting temperatures, the powder metallurgy process is often the practical method for processing. Materials that have high hardness and are used to form products with complex surfaces are frequently formed by the powder metallurgy process.

Parts produced through the use of powder metallurgy processing methods can be found in many areas of industry today. Some common parts that are produced via powder metallurgy include gears, bearings, rotors for pumps, cams, levers, magnets, and tool inserts (2). Gears are made by the powder metallurgy process, which, because it does not require machining of the gear teeth, greatly reduces the cost of gear manufacture. Figure 11.1 illustrates some of the products made by means of the powder metallurgy process.

11.2 POWDER METALLURGY PROCESS

The procedure for making parts via the PM process involves five steps:

1. Produce the powders.
2. Mix the powders.
3. Form the part.
4. Sinter the part.
5. Perform the secondary operations.

The first step is the forming of the powders, which can be accomplished via several different methods. The key is to get the proper size mesh (100–325 mesh), the proper size distribution, and the proper shape (spherical versus cylindrical). Care must be taken in the production of powders, since fine particles have a large surface area and can oxidize rapidly (explode) in the presence of air. The metals that oxidize rapidly, such as aluminum and magnesium, must be made into powders under carefully controlled conditions. Some of the powder production methods used follow.

Atomization: vaporizing the liquid metal and letting the small drops solidify into a spherical powder shape.

Reduction: reducing powder oxides to oxide metal if the metal has a high melting point.

Electrolytic deposition: making a powder deposit instead of a plating.

Mechanical crushing and grinding

Precipitation of powder from a solution

Figure 11.1 Typical products made by means of the powder metallurgy process. (Reprinted with permission from the Metal Powder Industries Federation, Princeton, New Jersey.)

The powder production method selected depends upon the required characteristics of the particular metal. Usually the powder is a raw material obtained from a supplier and is not one of the steps in the manufacturing process.

The mixing of the powders for the desired composition is necessary to ensure homogeneity of the composition. Lubricants, such as wax, are added to reduce the friction at the powder surfaces during compaction. The lubricant is removed via volatilization during the sintering operation.

The forming of the part can occur by pressing the powders either at room temperature or heated up. The pressures are relatively high, from 5 to 50 tons/in.^{2} (70-700 Mpa). If the temperature is elevated, the process is called HIP, for *hot isostatic pressing*. With the hot pressing, the parts are more than 99 percent dense. On the other hand, with cold forming, the density of the parts varies from 75 to 95 percent, depending upon the pressures used. Figure 11.2 illustrates some of the typical operations and tooling to form a "green" part, that is, the part before it is sintered to reach its desired mechanical properties.

The fourth step is the sintering of the part. If the part is formed by hot isostatic pressing, the sintering occurs during the forming operation. When cold forming is used, a separate operation is required for the sintering. The sintering is usually done by placing the parts on a metal conveyor or in a metal basket on the conveyor and slowly moving the parts through the furnace. The main considerations in the sintering operation are:

Temperature: The sintering temperature is approximately 70–80 percent of the melting temperature on the absolute scale.

Time: The sintering time varies from 10 to 480 min, with the higher times being for metals with high melting points.

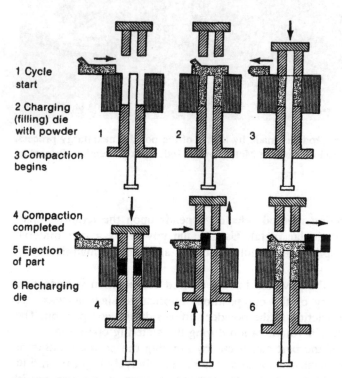

1 Cycle start

2 Charging (filling) die with powder

3 Compaction begins

4 Compaction completed

5 Ejection of part

6 Recharging die

Figure 11.2 Process operations and tooling for compacting powder. (Reprinted with permission from the Metal Powder Industries Federation, Princeton, New Jersey.)

Atmosphere: The atmosphere is reducing or inert, to avoid oxidation of metal; hydrogen and ammonia are gases that are often used.

The fifth step is the secondary operations that are necessary to finish the product. These can be grouped into the following two categories:

Manufacturing operations	Finishing operations
Oil impregnation	Heat treating
Forging (repressing)	Tumbling
Metal infiltration	Plating
Machining	Trimming

Table 11.1 summarizes the steps in the procedure for making parts via the powder metallurgy process. Machining is usually not required, but to reduce the complexity of the dies for forming it is usually easier and less expensive to drill holes after the part is made rather than to make metal cores for the dies to produce the holes during the forming of the part.

11.3 PROCESS ADVANTAGES, UNUSUAL PROPERTIES, AND PROCESS LIMITATIONS

The powder metallurgy process has unusual properties, since the density of the products can be varied via the amount of force applied during the forming stage. Some of the unique characteristics of powder metallurgy products follow.

1. Density can be varied, which permits weight reduction.
2. Strength and toughness both increase with increasing density, whereas for most materials toughness usually decreases when strength increases.
3. Alloy compositions can be made using metallics and nonmetallics that cannot otherwise be formed.
4. It is the best viable process for manufacturing parts from materials with poor workability, such as tungsten, beryllium, carbides, and ceramics (1), or with extremely high melting points, which restrict casting.
5. The part can be impregnated with oil for self-lubrication; this is used for self-lubricating gears.

Within the last ten years, the use of powder metallurgy processes to produce parts has increased rapidly. There is a good possibility that its use

Table 11.1 Steps in the Powder Metallurgy Process

Step	Description
1. Raw materials	Metal powders: Elemental or alloys Control size and particle distribution Additives: die lubricants, wax, graphite
2. Mixing	Mix powders and additives to obtain uniform composition
3. Compaction	Different compaction methods used: Hot compaction vs. cold compaction

1. Isostatic	1. Isostatic
2. Die compacting	2. Die compacting
3. Extrusion	3. Rolling
4. Spraying	4. Injection molding
5. Pressureless sintering	5. Slip casting

Step	Description
4. Sintering	Main process variables (cold compaction only): 1. Temperature 2. Time at temperature 3. Atmosphere/vacuum
5. Secondary operations	

Manufacturing	Finishing
Impregnation	Heat treating
Oil	Tumbling
Metal	Plating
Plastic	Trimming
Machining	Shot peening
Forging	Steam treating
Resintering	
Repressing	

in manufacturing in the coming years will continue to grow at a rate of 10–20 percent per year (2). Some of the process advantages of PM are:

1. The process has a high yield compared to the other primary processing methods, usually over 95 percent.
2. Complex and thin shapes can be produced relatively easily.
3. Few machining operations are required, for the part has a good finish.

4. High production levels and good reproducibility are typical of most manufacturing with PM parts.

Some of the process limitations are:

1. The tooling costs are high, because high pressures are involved and product shapes are intricate.
2. The material costs are high, because the powder form is more costly than bars, sheets, or slabs.
3. Mechanical properties (such as strength) are lower if the part is not 100 percent dense.
4. Solid state shrinkage will occur during sintering and cooling.
5. There are several design limitations, including:
   a. Generally small, intricate shapes are produced, with a maximum weight of 70 lb (30 kg).
   b. Uniform sections are preferred—grooves, re-entrant tapers, and threads are generally not possible, since powder cannot be compacted to uniform density.
   c. The minimum section thickness is 0.060 in. (1.5 mm).
   d. The length/diameter ratio should be under 5 for cylinders.
   e. Planar thin shapes, two-dimensional, are best (gears, knife blades, etc.)

11.4 DESIGN CONSIDERATIONS

There are two design calculations that frequently occur in the production of powder metallurgy parts. The first consideration is to determine the amount of materials required to make the part to a specific density in the green state (before sintering). The second is to determine the change in linear dimensions from the volume contraction during sintering.

The first design problem is that the mix ratio is in weight percent, whereas the part design involves the volume of the part (from the part shape) and the density of the mixture is not known. The calculations involve the basic relationship between mass and volume, that is, density. To illustrate the calculation procedure, an example problem will be utilized.

Example Problem 11.1. A part is to be made from three materials in the specified mix ratio. The part has a volume of 3 in.^{3}, and the green part is 80 percent dense. Determine the amounts of materials to be used, and estimate the mechanical strength of the part. The basic data is in Table 11.2.

Since the part is only 80 percent dense, the amount of void space is 20 percent, or 0.60 in.^{3}

Table 11.2 Material Property Data for Example Problem 11.1

Material	Density (d) (lb/in.^{3})	Weight fraction (WF) (decimal)	Material strength (psi)	Material cos ($/lb)
A	0.05	0.50	10,000	4.00
B	0.10	0.30	18,000	3.00
C	0.20	0.20	45,000	2.00

$$\text{volume} \times \text{percent void space} = \text{volume of void space}$$
$$3 \text{ in.}^3 \times 0.20 = 0.60 \text{ in.}^3$$

The total volume of the part can be written as:

$$V_T = V_A + V_B + V_C + \text{void space} \tag{11.1}$$

where

V_T = total volume of part
V_A = volume of part from material A
V_B = volume of part from material B
V_C = volume of part from material C
void space = volume of part not filled

   The total mass of the part can be written as the sum of the material masses:

$$M_T = M_A + M_B + M_C \tag{11.2}$$

where

M_T = total weight of part
M_A = weight of part due to material A
M_B = weight of part due to material B
M_C = weight of part due to material C

The void space does not contribute to the weight of the material and is not included in the calculation.

   The volume of the part can be determined from the individual volumes via the density and masses, since:

$$V_X = M_X/d_X \tag{11.3}$$

where

V_X = volume of material X in final component
M_X = mass of material X
d_X = density of material X

But the mass of material X is:

$$M_X = WF_X \times M \tag{11.4}$$

where

WF_X = weight fraction of material X
M = total mass of part M

If Eqs. (11.4), (11.3), and (11.2) are substituted into Eq. 11.1, the total mass can be found by:

$$M = \frac{V_T - \text{void space}}{WF_A/d_A + WF_B/d_B + WF_C/d_C} \tag{11.5}$$

For this problem, the total mass would be:

$$M = \frac{3.0 - 0.6}{0.5/0.05 + 0.30/0.10 + 0.20/0.20} = 0.171 \text{ lb}$$

The individual mass components can be found by multiplying the total mass by the weight fraction; thus,

$M_A = 0.50 \times 0.171 = 0.086$ lb

$M_B = 0.30 \times 0.171 = 0.051$ lb

$M_C = 0.20 \times 0.171 = 0.034$ lb

The density of the part is the total mass divided by the total volume, or:

$d_T = 0.171$ lb/3.0 in.^{3}

$= 0.057$ lb/in.^{3}

The total material cost would be the sum of the individual mass components times their respective unit costs (Table 11.2) as indicated by:

$C_u = C_A \times M_A + C_B \times M_B + C_C \times M_C$

$C_u = \$4.00 \times 0.086 + \$3.00 \times 0.051$ (11.5a)

$+ \$2.00 \times 0.034 = \$.565$

The strength of the component can be estimated by means of the volume fraction of the components and can be expressed as:

$$\sigma = \sigma_A \times VF_A + \sigma_B \times VF_B + \sigma_C \times VF_C \tag{11.6}$$

where

$$VF_A = \frac{M_A/d_A}{V_T} = \text{volume fraction of } A$$

$$VF_B = \frac{M_B/d_B}{V_T} = \text{volume fraction of } B$$

$$VF_C = \frac{M_C/d_C}{V_T} = \text{volume fraction of } C \tag{11.7}$$

Thus,

$$\sigma = 10,000 \times \frac{0.086/0.05}{3.0} + 18,000 \times \frac{0.051/0.10}{3.0} + 45,000 \times \frac{0.034/0.20}{3.0}$$

$$= 5,733 + 3,060 + 2,550$$

$$= 11,343 \text{ psi}$$

The second design problem concerns the dimensions of the part after the sintering operation.

Example Problem 11.2. If the density of the part is increased to 95 percent from 80 percent and the original part shape was 3 in. × 1 in. × 1 in., what are the final part dimensions?

The percentage refers to the volume percent, not the linear percentage. The final volume is smaller, because the density has increased; the final volume is:

$$V_{Final} = V_{Initial} \times d_{Initial}/d_{Final} \tag{11.8}$$

where the initial and final percentage densities are given. For Example Problem 11.2, the value would be:

$$V_{Final} = 3.0 \text{ in.}^3 \times 80/95$$

$$= 2.52 \text{ in.}^3$$

The percentage of the original length is a function of the cube root and can be expressed as:

$$L_{Final} = L_{Initial} \times (d_{Initial}/d_{Final})^{0.333} \tag{11.9}$$

For the example problem, the final length values would be:

$$L_{Final} = L_{Initial} \times (0.80/0.95)^{0.33}$$

$$= L_{Initial} \times .944$$

Thus, the final dimensions of the 3 × 1 × 1 in. would be 2.833 × 0.944 × 0.944 in. The amount of shrinkage can be determined by:

linear shrinkage (decimal) = $1.00 - (d_{Initial}/d_{Final})^{0.333}$ (11.10)

For this particular example problem, the value is:

linear shrinkage = $1.00 - 0.944$
$$= 0.056, \text{ or } 5.6\%$$

The design engineer must accommodate the shrinkage in the design so the part will have the correct final dimensions. For example, if the final part was to have the dimensions of 3 in. $\times$ 1 in. $\times$ 1 in., the initial dimensions would be divided by the factor 0.944, and would become 3.178 in. $\times$ 1.059 in. $\times$ 1.059 in. This assumes that the shrinkage would be equal in all directions, an assumption that does not always hold. This shrinkage is due to the density change and is not a result of the thermal contraction during the cooling of the part.

11.5 SUMMARY

Powder metallurgy is a "near-net-shape" process that is somewhat related to both casting and forging processes. It has a higher yield that either process and is generally used to produce small, complex-shaped symmetrical parts. It permits utilization of some material compositions that cannot normally be processed by casting, forging, or machining. The calculation of the material mixes to produce the desired compositions is unique to powder metallurgy processes. The shrinkage calculations are to determine the shrinkage that occurs as a result of density changes during sintering. The tooling and material costs are relatively high, but better tolerances and surface finish reduce the number of secondary operations required. Materials with high melting temperatures and/or high hardness are difficult to form by the other traditional processes and frequently are formed by the powder metallurgy process.

11.6 EVALUATIVE QUESTIONS

1. What are the five steps in the powder metallurgy process?

2. In the sintering operation, describe the three main considerations.

3. Discuss some of the advantages of powder metallurgy.

4. A powder metallurgy product is made of Al (90 percent) by weight, which has a density of 2.69 g/cm^3 and Fe (10 percent), which has a density of 7.87 g/cm^3. The final volume of the product is 2 cm^3.

a. What is the theoretical density of the product?

b. What is the weight of the mix to give the 2-cm³ volume?

5. A part has a green density of 0.8 g/cm³ and a sintered density of 0.95 g/cm³. What linear shrinkage occurred during sintering? (Assume a cube with a 1-cm edge)

6. A powder metallurgy part is to be made from the following materials. The final part will be only 95 percent dense and have a total volume of 4 in.³. The void spaces will be used for oil impregnation. The material information is:

Material	Density (lb/in.³)	Weight (%)	YS (kpsi)	Cost ($/lb)
D	0.10	40	20	6:00
E	0.20	50	30	4.00
F	0.25	10	80	3.00

a. Determine the weight of each of the components to be mixed and the total weight.

b. Estimate the tensile strength of the part if the properties vary according to the volume fraction.

c. Determine the material cost of the sintered part.

7. A powder metallurgy part is to be made from the following materials. The final part will have a volume of 2.5 in.³ and will be 90 percent dense. The material information is:

Material	Density (lb/in.³)	Weight (%)	YS (kpsi)	Cost ($/lb)
X1	0.10	30	20	1.5
X2	0.15	25	70	2.5
X3	0.25	45	50	1.0

a. Determine the weight of each of the three materials, in pounds to be mixed together, to make the 2.5-in.³ part of 90 percent density.

b. Determine the material cost of the part.

c. Determine the density of the part, in lb/in.³.

d. Estimate the yield strength of the part if the properties vary according to the volume fraction in kpsi.

REFERENCES

1. Boyer, H. E., and Gall, T. L., eds. The Metals Handbook, Desk Edition, 1985, American Society for Metals, Chapter 25, "Powder Metallurgy," pp. 25-1–25-24.
2. Alting, Leo. Manufacturing Engineering Processes, 2nd ed., 1994, Marcel Dekker, New York, pp. 282–300.
3. Metal Powder Industries Federation, "Powder Metallurgy Design Solutions," 1993, pp. 3, 19.

BIBLIOGRAPHY

The following organization has considerable information on powder metallurgy. One of their basic introductory pamphlets, "Design Solutions," is quite useful for students and engineers.

Metal Powder Industries Federation
105 College Road East
Princeton, NJ 08540-6692
Tel: 6099-452-7700

INTERNET SOURCES

Metal Powder Industries Federation: *http://www.mpif.org*
http://www.epma.com/process/welcome.html
http://www.met.mat.ethz.ch/brochure.html

12
Bulk Deformation Processing

12.1 INTRODUCTION

Deformation processing, also called metalworking or metalforming, is a process where conservation of mass and extensive plastic deformation occur. There are several types of classification systems used for deformation processing (1), such as type of workpiece (bulk or sheet), temperature of working (hot, warm, or cold), mode of deformation (steady state, non–steady state, or mixed), and system of stresses (compression, tension, bending, torsion, shear, or combined tension and compression). Deformation processing was traditionally classified by the type of operation (primary or secondary). The primary processing produced near-net shapes that required further processing (such as bars, rods, and slabs), and the secondary processing produced the final shapes. The problem was that some operations were primary for some products but secondary for other products. The method most used classifies processes according to the type of workpiece. The two main areas are bulk deformation and sheet metal deformation. These two classes are described in Table 12.1. Some additional comments follow.

Bulk deformation generally has the starting material in the form of semifinished shapes that have a high volume-to-surface-area ratio, or high modulus. The bulk deformation increases the surface area and thus decreases the value of the modulus. Generally, deformation is three dimensional and elastic recovery is insignificant.

Sheetmetal processes have rolled sheet as the starting material, and these have a low volume-to-surface-area ratio, or low modulus. The sheet metal forming or sheetforming processes change the shape but do not greatly alter the value of the modulus, because the change in thickness is small. Generally, sheet metal forming processes have two-dimensional deformation, and the elastic recovery is usually significant.

Table 12.1 Major Deformation Categories, Discriminating Factors, and Typical Processes

Bulk deformation processes				Sheetmetal processes				
A. Discriminating Factors (Decreasing pressure-multiplying factor) (3D complex → 2D simple)				A. Discriminating Factors (General forming complexity via surface affected) (Large thickness change → little change) (cutting)				
B. Typical Processes: Major Classifications and Subgroups				B. Typical Processes: Major Classifications and Subgroups				
Forging	Extrusion	Wire drawing	Rolling	Spinning	Stretch drawing	Deep drawing	Bending	Shearing
Open-die	Forward	Wire	Flat	Conventional	Stretch Forming	Circular	Press	Guillotine
Impression-die	Backward	Tube	Shape	Shear		Rectangular	Roll	Blanking
Closed-die			Ring	Tube			Air	Piercing
Swaging								

12.2 BULK DEFORMATION PROCESSES

The most common bulk deformation processes are forging, extrusion, wire drawing, and rolling, and the following brief descriptions (2) of each process are presented.

Forging is defined as the controlled plastic deformation of metals into a desired shape by means of applied pressure. This is done by presses, impact hammers, dies, or other related machinery. There are three types of forging: *Open-die* forging involves the compression of material between two flat or shaped dies where the die does not constrict material flow. This process is generally used for large parts with simple shapes and small quantities, but is not limited to such. For *impression-die* forging, the die contains recesses that allow the workpiece to be formed into three-dimensional shapes with complex surfaces. As in open-die forging, the flow of material is not completely restricted, and excess, called *flash*, occurs. This process can produce parts of various sizes and quantities and is the most common type of forging. Some parts produced by impression-die forging with loose tooling are illustrated (3) in Fig. 12.1. Closed-die forging is similar to impression-die forging; however, the workpiece is completely enclosed in the die, producing a part that is near-net shape and needs little secondary operations. This is normally used for small parts and high production quantities, such as the various coins—pennies, nickels, dimes, quarters, and half-dollars.

Extrusion is defined as a plastic deformation process in which metal or plastic is forced to flow through the shaped orifice of a die. It is normally used to produce long parts of constant cross-sectional area. Parts produced by extrusion generally have close tolerances and good surface finishes. Some of the shapes produced by forward, backward, and combined forward and backward extrusion are illustrated in Fig. 12.2.

Wire drawing involves pulling metal through a die to produce a bar, a rod, or a wire. The parts produced by this method are usually longer and smaller in cross-sectional area than those produced by extrusion. Wire drawing involves pulling the metal, whereas extrusion generally involves pushing the metal.

Rolling involves the continuous forcing of metal between two rotating rolls on opposing sides of the workpiece. The parts produced have a uniform cross-sectional area. They are generally longer than those produced by extrusion and larger than those produced by wire drawing. Structural shapes, such as I-beams, and railroad track rail are produced by rolling.

The bulk deformation processes compete with castings for the production of many products. The advantages of bulk deformation over casting follow.

1. *Improved microstructural quality*: Wrought products have more homogeneous compositions and grain size as well as less porosity than cast-

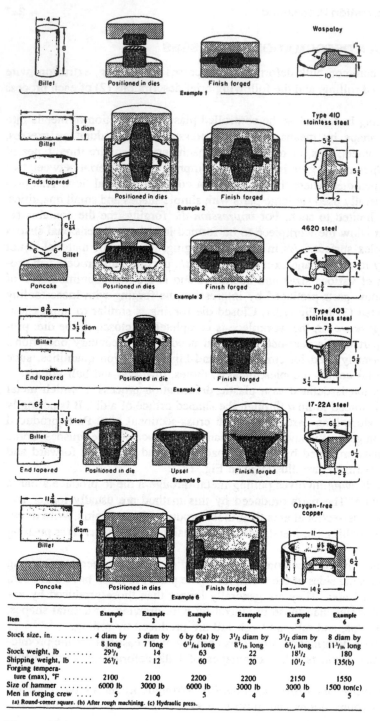

Item	Example 1	Example 2	Example 3	Example 4	Example 5	Example 6
Stock size, in.	4 diam by 8 long	3 diam by 7 long	6 by 6(a) by 6^{11}/$_{64}$ long	3^1/$_2$ diam by 8^1/$_{16}$ long	3^1/$_2$ diam by 6^1/$_4$ long	8 diam by 11^1/$_{16}$ long
Stock weight, lb	29^1/$_4$	14	63	22	18^1/$_2$	180
Shipping weight, lb	26^1/$_4$	12	60	20	10^1/$_2$	135(b)
Forging temperature (max), °F	2100	2100	2200	2200	2150	1550
Size of hammer	6000 lb	3000 lb	6000 lb	3000 lb	3000 lb	1500 ton(c)
Men in forging crew	5	4	5	4	4	5

(a) Round-corner square. (b) After rough machining. (c) Hydraulic press.

Figure 12.1 Examples of impression-die forging. (Reproduced with permission of ASM International from Ref. 13.)

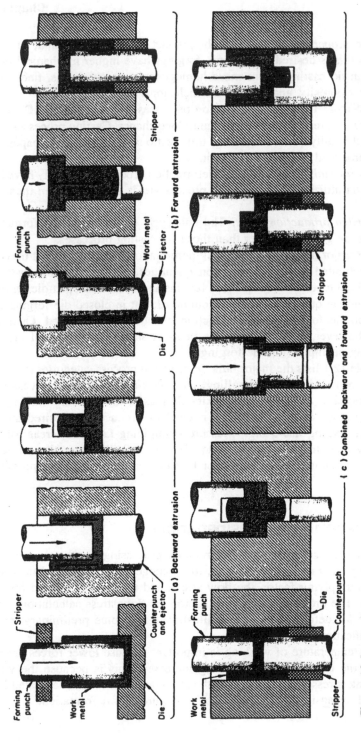

Figure 12.2 Examples of extrusion: (a) backward, (b) forward, and (c) combined forward and backward. (Reproduced with permission of ASM International from Ref. 13.)

Forming punch

Stripper

Work metal

Die

Counterpunch and ejector

(a) Backward extrusion

Forming punch

Work metal

Die

Stripper

Ejector

(b) Forward extrusion

Work metal

Forming punch

Die

Counterpunch

Stripper

(c) Combined backward and forward extrusion

ings. The properties of the wrought product may be directional; in some instances this is desired. In general, forgings have higher mechanical properties than do castings from more homogeneous compositions, finer grain size, lower porosity, and less severe inclusions.

2. *Shape*: The bulk deformation processes permit complex 2D shapes (internal and external) to be produced in long shapes. Casting processes have extreme difficulty producing long, thin sections. Complex 3D shapes can also be produced externally in forgings.

3. *Material utilization*: Bulk deformation processes generally have better material utilization, because the process yields are higher than those for castings.

4. *High production rates*: The unit cycle times for bulk deformation processes are generally lower than those for casting processes.

The major disadvantage of bulk deformation processes is that they are generally limited to high production quantities, because tooling costs for deformation processes are high. In a few instances, such as open-die forging, low production quantities can be economical; but in closed-die forging and impression-die forging, high production quantities are required. Open-die forging, however, does not have the near-net shape capabilities of closed-die forging or impression-die forging.

Table 12.1 lists the major bulk deformation process categories (forging, extrusion, wire drawing, and rolling) and gives a few specific processes in these categories. As shape complexity decreases, the pressure-multiplying factor decreases and less force is needed to produce the desired shape. Closed-die forging has a high pressure-multiplying factor, whereas rolling has a low pressure-multiplying factor. Figure 12.3 illustrates the variation of the pressure-multiplying factor for frictionless, plain strain compression.

12.3 DEFORMATION CONSIDERATIONS

Some of the considerations in deformation processing are the temperature of working, the rate of deformation, the homogeneity of material flow, the pressure-multiplying factor, the flow stress, and the forces required for deformation. The pressure-multiplying factor, the flow stress calculations, and the forces for deformation will be illustrated, but some preliminary terms and classifications are required.

The temperature of working is important because lower forces are required as the temperature of working increases. This is because the yield and ultimate stresses are reduced as the slip planes are further separated due to thermal expansion. Thus, less force is required for deformation.

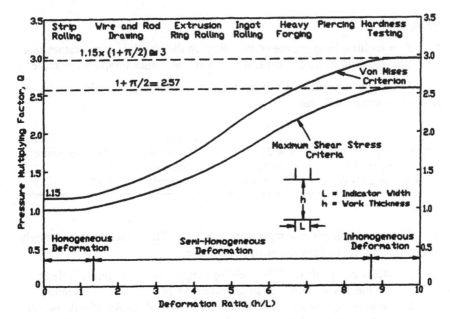

Figure 12.3 Variations of the pressure-multiplying factor for deformation mode for frictionless plain strain indentation. (Adapted from Ref. 11.)

There are generally three classifications of processes based on the temperature of working:

Hot working (HW): the mechanical working of the material above the recrystallization temperature

Cold working (CW): the mechanical working of the material below the recrystallization temperature

Warm working (WW): the cold working of the material at a temperature elevated above room temperature but below the recrystallization temperature

The type of working is based upon the recrystallization temperature of the material being worked. The recrystallization temperature (T_R) is defined as the temperature at which 95 percent of the recrystallization is complete in 1 hr.

An approximate range for the recrystallization temperature is:

$$T_R = (0.33 - 0.50)T_m$$

where

T_m = melting temperature of the alloy on the absolute temperature scale
T_R = recrystallization temperature on the absolute temperature scale

Thus, the temperature ranges for the working of metals can be expressed approximately as:

Cold working (CW)	$T < T_R$	Generally at room temperature
Hot working (HW)	$T \geq T_R$	Generally $0.5-0.8T_m$
Warm working (WW)	$T_{room} < T < T_R$	Generally $0.1-0.5T_R$

The recrystallization temperature can vary. Three factors that strongly influence the recrystallization temperature are:

1. *Alloy composition*: The melting temperature is strongly influenced by the alloy composition as indicated on the phase diagrams.
2. *Amount of strain*: A minimum amount of strain (5–10 percent CW) is required for recrystallization to occur; the more the strain, the lower the recrystallization temperature.
3. *Phase transformations*: When a material undergoes a phase transformation, the new phase will be nucleated and grow. In the instances where the phase transformation temperature is below the estimated recrystallization temperature, the recrystallization will occur at the transformation temperature.

Three stages have been observed in the recrystallization process:

1. *Recovery stage*: The removal of point defects via diffusion; this increases conductivity but has only a small effect upon strength.
2. *Recrystallization stage*: The formation of "strain-free" grains, the removal of the line defects or dislocations. Generally, strength is greatly decreased and ductility greatly increased.
3. *Grain growth stage*: The growth of grains results in fewer grains and less grain boundaries. This results in a small decrease in strength and a small increase in ductility.

Deformation processes are also classified with respect to the rate of deformation and the homogeneity of flow. With respect to deformation, processes are classified as:

1. *Steady state deformation*: Processes in which all of the part is subject to the same mode of deformation; examples are rolling, drawing, and some extrusion processes.

2. *Non–steady state deformation*: Processes in which geometry continually changes, the mode of deformation changes during the process, and shape tends to be complex; examples are shape forging and complex extrusions.

The homogeneity of flow results in three classifications:

1. *Homogeneous deformation*: The deformation is uniform throughout the thickness of the material; the pressure-multiplying factor is unity (or 1.15).
2. *Semihomogeneous deformation*: Nonuniform deformation occurs in the material. The center and surface deformations are different.
3. *Inhomogeneous deformation*: Nonuniform deformation in which deformation only at the surface, not at the center.

12.4 FORCE DETERMINATION

One of the considerations in bulk deformation problems is the size of the press needed for deformation. This requires the determination of the amount of force needed to cause the required deformation, which in turn requires the flow stresses and pressures to form the product. There are several advanced methods for the calculations, including finite-element and finite-difference methods, matrix methods, weighted residuals, slab method, slip line, upper-bound, and uniform energy. For precise calculations, these methods should be investigated for the most appropriate procedure for the specific conditions. The following procedure is a modification of procedures (1,4,5,6) in various textbooks and is only for estimating purposes. *Metalworking Science* and *Engineering* by Mielnik (1) presents more detailed methods for making calculations.

1. Determine the true strain imparted to the material and the true strain rate. The true strain is determined by:

$$\varepsilon = \ln(l_f/l_o) \tag{12.1}$$

and the true strain rate by:

$$\varepsilon = v/l \tag{12.2}$$

2. Determine the relevant flow stress $[\sigma_{(f \, or \, fm)}]$. The three different expressions for flow stress follow.

   a. cold-worked materials (non–steady state deformation):

$$\sigma_f = K\varepsilon^n \tag{12.3}$$

   where K and n are constants as found in Table 12.2.

Table 12.2 Flow Stress Equation Constants

Material designation	Temperature (°C)	Hot working C (MPa)	(kpsi)	m	Cold working K (MPa)	(kpsi)	n
Steels							
1008 carbon steel	Room temp.	—	—	—	660	95	0.24
1015 carbon steel	Room temp.	—	—	—	786	114	0.10
	800	148	21.5	0.10			
	1000	130	18.8	0.06			
	1200	47	6.8	0.17			
1045 carbon steel	Room temp.	—	—	—	1020	148	0.11
	900	170	24.6	0.11			
	1000	109	15.8	0.16			
	1100	73	10.6	0.18			
302 stainless steel	Room temp.	—	—	—	1520	220	0.60
	1000	177	25.7	0.102			
	1100	130	18.9	0.108			
	1200	96	13.9	0.12			
410 stainless steel	Room temp.	—	—	—	950	138	0.09
Copper alloys							
Cu (99.9%)	Room temp.	—	—	—	436	63.3	0.33
	600	88	12.7	0.06			
	750	52	7.6	0.10			
	900	32	4.7	0.13			
Cartridge brass	Room temp.	—	—	—	441	64	0.41
	600	110	16	0.20			
	800	49	7.1	0.14			
Aluminum alloys							
1100	Room temp.	—	—	—	119	17.3	0.30
	200	80	11.7	0.07			
	400	38	4.4	0.13			
	500	14	2.1	0.23			
2017	Room temp.	—	—	—	311	45.2	0.18
	200	238	34.5	0.01			
	400	102	14.8	0.11			
	500	40	5.8	0.13			
5052	Room temp.	—	—	—	203	29.4	0.13
	340	61	8.9	0.07			
	440	38	5.6	0.13			

Table 12.2 Continued

Material designation	Temperature (°C)	Hot working			Cold working		
		C			K		
		(MPa)	(kpsi)	m	(MPa)	(kpsi)	n
Titanium alloys							
Ti (99.9%)	Room temp.	—	—	—	640	92.8	0.03
	200	420	61	0.046			
	400	274	40	0.074			
	600	174	25	0.10			
	800	88	12.8	0.17			
	1000	21	3.0	0.39			
Ti-5Al-4V		—	—	—	1411	205	0.018
	200	1041	151	0.021			
	400	875	127	0.022			
	600	648	94	0.064			
	800	353	51	0.146			
	1000	66	9.5	0.131			

From Ref. 16.

    b. Hot-worked materials:

$$\sigma_f = C\dot{\varepsilon}^m \tag{12.4}$$

where C and m are constants as found in Table 12.2.

    c. Steady state deformation, or mean flow stress (cold worked)

$$\sigma_{fm} = K\varepsilon^n/(n + 1) \tag{12.5}$$

where K and n are constants as found in Table 12.2.

    d. When no formulas are available for the determination of the flow stress or mean flow stress, the ultimate tensile strength (UTS) can be used as an approximation. The UTS is a high estimate, for the flow stress is usually between the yield and ultimate values, but it would be better to overestimate rather than underestimate the forces required for deformation. The flow stress, however, is closer in value to the UTS value than to the yield stress, and, therefore, an average value would tend to be low.

3. Find the pressure-multiplying factor (Q): There are several different methods for estimating the pressure-multiplying factor. One approach (4) is to use tabular values; this has been done for impression-die forging. Some of the values used are presented in Table 12.3.

Table 12.3 Pressure-Multiplying Factors (*Q*) for Impression-Die Forging

Part description	Shape description	Pressure-multiplying factor range
Die forging	Simple shapes, no flash	3–5
	Simple shapes, flash	5–8
	Complex shapes,[a] flash	8–12

[a]Tall ribs, thin webs as part of shape.
From Refs 4, 14, 15.

Formulas have been developed for different types of loading, such as plain strain, axisymmetric compression, rolling, and hardness testing, as a function of the coefficient of friction and forming shape parameters. These expressions are approximations, and figures for the expressions are in texts such as Mielnik (1), Schey (4), Kalpakjian (6), Alexander et al. (7), and Edwards and Endean (8). The expressions are presented in Table 12.4 for various forming processes.

4. Determine the average pressure (*P_a*): The average pressure is calculated from the flow stress and the pressure-multiplying factor by:

$$P_a = Q\sigma_{(f\,or\,fm)} \tag{12.6}$$

5. Determine the force required (*F*): The force required for the deformation is calculated from the average pressure and the die opening area or final shape area for flat parts. For parts with complex shapes, such as forgings, the area is the die opening area, and the pressure-multiplying factor is usually high:

$$F = P_a A \tag{12.7}$$

Combining Eqs. (12.6) and (12.7), the force can be expressed as:

$$F = Q \times \sigma_{(f\,or\,fm)} \times A \tag{12.8}$$

Since the forces can be quite high, forging is often done in steps, to reduce the pressure-multiplying factor.

Example Problem 12.1. A square bar, 0.25 × 0.25 in. in cross-section and 4 in. long is to be flattened into a section that is only 0.05 in. thick and will remain 4 in. in length. The press velocity is 50 in./min; the material is copper; and the coefficient of friction (μ) is 0.10. A sketch of the initial and final states of the bar is shown in Figure 12.4. The following questions are to be answered.

Table 12.4 Pressure-Multiplying Factors (Q) for Bulk Deformation Processes

Application description	Expression	Restriction	Terms
Plane strain compression	$Q = 1 + \mu L/2h$	$\mu L/h \leq 1$	μ = coefficient of friction L = width dimension (changing) h = thickness dimension (changing) (sticking, no lubrication)
	$Q = 1 + L/4h$	$\mu L/h > 1$	
Axisymmetric compression	$Q = (2/\alpha^2)(e^\alpha - 1 - \alpha)$	$\alpha \leq 1$	$\alpha = \mu d/h$ d = diameter (changing) h = thickness (changing) (sticking, no lubrication)
	$Q = 1 + \dfrac{d/h}{3\sqrt{3}}$	$\alpha > 1$	
Rolling	$Q = 1 + (\mu/\sqrt{2})(r/h)^{1/2}(b/h)^{1/2}$ (Ref. 8)		μ = coefficient of friction h = thickness after rolling $h + 2b$ = thickness before rolling r = radius of roll
Extrusion	$Q = [1 + \mu \cot \alpha]\phi \ln R$ $\phi = 0.88 + 0.12h/L$		μ = coefficient of friction ϕ = inhomogeneity factor h = mean (average) diameter L = die contact length $= (d1 - d2)/2 \sin \alpha$ α = half-angle of draw die R = area ratio, A1/A2 A1, $d1$ = initial area, diameter A2, $d2$ = final area, diameter
Hardness testing	$Q = 3$		

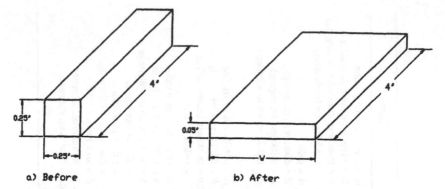

a) Before   b) After

Figure 12.4 Sketch of bar before and after flattening by plain strain deformation.

a.   What is the final width (w) of the material?
b.   What is the engineering strain in the thickness direction?
c.   What is the true strain in the thickness direction?
d.   What is the engineering strain rate?
e.   What is the true strain rate?
f.   Estimate the pressure-multiplying factor.
g.   What is the flow stress (psi) if cold working is performed?
h.   What is the flow stress (psi) if hot working is done at 600°C?
i.   What is the mean flow stress (psi) if cold working is done?
j.   If hot working is the process used, what press force is needed for the flattening operation?

Solution:

a.   The volume of the material is conserved, so the final width can be determined from:

$$4 \text{ in.} \times 0.25 \text{ in.} \times 0.25 \text{ in.} = 4 \text{ in.} \times 0.05 \text{ in.} \times w$$

Thus, $w = 1.25$ in.

b.   The engineering strain can be found from Eq. (3.4); that is:

$$e = \delta l / l_o = (0.05 - 0.25)/0.25 = -0.80$$

c.   The true strain can be found from Eq. (12.1); that is:

$$\varepsilon = \ln(l_f/l_o) = \ln(0.05/0.25) = -1.60$$

d.   The engineering strain rate can be found from Eq. (3.8); that is:

$$e = v/l_o = \frac{50 \text{ in./min} \times 1 \text{ min}/60 \text{ sec}}{0.25 \text{ in.}} = 3.33 \text{ sec}^{-1}$$

e. The true strain rate can be found from Eq. (12.2); that is:

$$\dot{\varepsilon} = v/l = \frac{50 \text{ in./min} \times 1 \text{ min/60 sec}}{0.05 \text{ in.}} = 16.67 \text{ sec}^{-1}$$

f. The pressure-multiplying factor can be estimated from the data in Table 12.4. The value of $\mu L/h$ must be determined first:

$$\mu L/h = 0.10 \times 1.25 \text{ in./0.05 in.} = 2.5 > 1$$

Since the ratio is greater than 1, this implies that sticking will occur, and the expression that should be used for Q is:

$$Q = 1 + L/4h = 1 + 1.25/(4 \times 0.05) = 7.25$$

g. If cold working is performed and copper is used, the values of K and n from Table 12.2 in Eq. (12.3) result in:

$$\sigma_f = K\varepsilon^n = 63.3 \text{ kpsi} \times 1.6^{0.33}$$
$$= 73.9 \text{ kpsi} = 73,900 \text{ psi}$$

Note. Although the strain was -1.6, the absolute value of 1.6 is entered in the equation. The minus sign implies compression instead of tension, but the formula is for either.

h. If hot working is done at 600°C on the copper material, the values of C and m in Table 12.2 with Eq. (12.4) result in:

$$\sigma_f = C\dot{\varepsilon}^m = 12.7 \text{ kpsi} \times 16.67^{0.06}$$
$$= 15.0 \text{ kpsi} = 15,000 \text{ psi}$$

i. If the mean flow stress expression of Eq. 12.5 is used, the flow stress (using the data in Table 12.2) is:

$$\sigma_{fm} = K\varepsilon^n/(n + 1) = 63.3 \text{ kpsi} \times 1.6^{0.33}/(1 + 0.33)$$
$$= 55.6 \text{ kpsi} = 55,600 \text{ psi}$$

j. If hot working is used, and the flow stress and pressure-multiplying factor have been calculated, the surface area is the only additional information needed. The force is a large number, so tons are the units usually used in the United States:

$$A_f = 4 \text{ in.} \times 1.25 \text{ in.} = 5 \text{ in.}^2$$

Thus,

$$F = 7.25 \times 15,000 \text{ psi} \times 5 \text{ in.}^2 \times 1 \text{ ton/2000 lb}$$
$$= 272 \text{ tons}$$

These calculations illustrate how the formulas are to be applied to determine the press force requirements. The problem also indicates the effects temperature has upon the flow stress and the lower forces required in hot working. More detailed data can be found in Schey (4,5) and more theoretical calculations in Mielnik (1) and Alexander et al. (7).

12.5 FORGING COST ESTIMATING

It is difficult to estimate the cost for forgings, but a novel approach was developed by Knight and Poli (9). This approach was developed to provide relative costs of producing parts with the same function, but with different design features. The approach is based on a group technology code for classifying part shapes and then determining the relative material, die, and operating costs. An explanation of the system developed has also been presented by Ludema, Caddell, and Atkins (10). The cost of a forging can be determined as:

$$K_T = K_m + K_p + K_d \tag{12.9}$$

where

K_T = total cost of the part

K_m = material costs (including scrap)

K_p = production costs (capital, direct labor, finance, setup, handling, overhead)

K_d = die costs

The basis of this system is first to determine the relative cost with respect to a reference part, and then to use this ratio to estimate the cost of the new part. The reference part is a basic disk shape from carbon steel. The parameters for this approach are:

K_{TO} = cost of reference part

K_{MR} = material cost ratio of part of interest to reference part; that is, K_m/K_{MO}

K_{PR} = operating cost ratio of part of interest to reference part; that is, K_p/K_{PO}

K_{DR} = die cost ratio of part of interest to reference part; that is, K_d/K_{DO}

K_{MO} = ratio of material costs to total costs for reference part; = $A \times K_{TO}$

K_{PO} = ratio of production costs to total costs for reference part; = $B \times K_{TO}$

K_{DO} = ratio of die costs to total costs for reference part;

$\quad = C \times K_{TO}$

but the sum of the cost fractions must be unity; that is:

$$A + B + C = 1.0 \tag{12.10}$$

Using the preceding definitions and Eq. (12.10) in Eq. (12.9), the ratio of K_T/K_{TO} can be obtained:

$$K_T = K_m + K_p + K_d \tag{12.9}$$

$$= (K_{MR} \times K_{MO}) + (K_{PR} \times K_{PO}) + (K_{DR} \times K_{DO})$$

$$= (K_{MR} \times A \times K_{TO}) + (K_{PR} \times B \times K_{TO}) + (K_{DR} \times C \times K_{TO})$$

Dividing by K_{TO}, one obtains:

$$K_T/K_{TO} = (A \times K_{MR}) + (B \times K_{PR}) + (C \times K_{DR}) \tag{12.11}$$

The total cost, K_T, can be determined from the ratio obtained in Eq. (12.11) and the cost of the reference part, K_{TO}.

The values of K_{MR}, K_{PR}, and K_{DR} can be high, since the reference part is mild carbon steel. The use of the expressions can be illustrated with an example problem.

Example Problem 12.2. A forging shop has a cost ratio of 20 percent for materials, 50 percent for production costs, and 30 percent for die costs. For a newly designed part, the relative material cost is 2, the relative production cost is 6, and the relative die cost is 15. If the reference part costs $4.00, what is the expected cost of the new part?

$$K_T/K_{TO} = 0.20 \times 2 + 0.50 \times 6 + 0.30 \times 15$$

$$= 7.90$$

$$K_T = 7.90 \times K_{TO}$$

$$= 7.90 \times 4$$

$$= \$31.60$$

The key is in the determining of the specific values for the relative material cost, relative production cost, and relative die cost. The references by Knight and Poli (9) and Ludema et al. (10) give more details on these calculations.

12.6 SUMMARY

Deformation processing is divided into bulk deformation and sheet metal working. The bulk deformation processes have the advantages of improved microstructure, better material utilization, high production rates, and the ability to produce long, complex two-dimensional shapes over castings. The two categories of hot and cold working are defined with respect to the recrystallization temperature. A procedure for calculating the press size force requirements was presented and illustrated with an example problem. The use of the relative costing approach for design evaluation was presented.

12.7 EVALUATIVE QUESTIONS

1. What are the two primary classifications of deformation processing, and how are they different (see Table 12.1)?

2. What are the advantages of bulk deformation processing over metal casting? What are the advantages of metal casting over bulk deformation processing?

3. What are the three classifications of working based upon recrystallization temperature? What is the recrystallization temperature? (See Ref. 12 for melting temperatures.)

4. What are the stages of recrystallization?

5. Estimate the recrystallization temperatures for pure copper, aluminum, lead, and iron and an AISI/SAE 1020 steel.

6. Repeat the calculations in Example Problem 12.1, but use a lubricant with a coefficient of friction of 0.05 and a starting material with a size of 0.20 in. × 0.20 in. × 4 in. and flatten it to a thickness of 0.05 in. × 4 in. × w.

7. Repeat the calculations in Example Problem 12.1, but use 5052 aluminum instead of copper. Use the same temperature when considering hot working.

8. A cylindrical sample with an initial diameter of 0.25 in. and a height of 0.20 in. is upset in an open-die forge to a height of 0.05 in. and remains circular in cross section. A lubricant is used with a coefficient of friction of 0.10, and the press velocity is 60 ft/min. Assume that cartridge brass is the material used and answer the following questions:
   a. What is the engineering strain in the height direction?
   b. What is the true strain in the height direction?
   c. What is the true strain in the radial direction?
   d. What is the pressure-multiplying factor?
   e. What is the mean flow stress (psi)?
   f. What is the flow stress if the metal temperature is 600°C?

g. If the calculated flow stress is 60,000 psi, what force in tons is needed to make the part?

h. If the material was cold worked, what is the expected BHN of the material based upon the flow stress for cold working?

9. The forge shop has a cost ratio of 40 percent for materials, 25 percent for production costs, and 35 percent for die costs. For a newly designed part, the relative material cost is 0.50, the relative production cost is 5, and the relative die cost is 12. If the reference part cost is $6.00, what is the expected manufacturing cost for the new part? ($33.90)

10. A square bar, 5 mm × 5 mm in cross section and 100 mm long is to be flattened into a section that is only 2 mm thick and remains 100 mm in length. The press velocity is 2 m/min. The coefficient of friction is 0.10, with a lubricant. The value for K is 620 MPa, and n is 0.18. The value of C is 120 MPa and m is 0.10 at 1000°C.

a. Make a sketch of the bar before and after deformation.
b. What is the engineering strain in the height direction?
c. What is the true strain in the height direction?
d. What is the true strain rate?
e. What is the pressure-multiplying factor?
f. What is the mean flow stress (MPa)?
g. What is the flow stress if the part is formed at 1000°C (MPa)?
h. What is the force required (newtons) if the part is formed at 1000°C?
(b. −0.6 c. −0.916 d. 16.67 sec^{-1} e. 1.312 f. 517 MPa g. 158.5 MPa h. 260 kN (259,500 N))

11. A cylindrical sample with an initial diameter of 2 in. and a height of 6 in. is upset in an open-die forge to a height of 1 in. and remains circular in cross section. A lubricant is used with a coefficient of friction of 0.10, and the press velocity is 60 ft/min. The material is a 1015 carbon steel at 800°C, and the value of C (MPa) is 148 and m is 0.10.

a. What is the engineering strain in the height direction?
b. What is the true strain in the height direction?
c. What is the pressure-multiplying factor?
d. What is the flow stress in psi?
e. What force is required in tons?
(a. −0.833 b. −1.79 c. 1.185 d. 27,500 psi e. 307 tons)

REFERENCES

1. Mielnik, E. M. Metalworking Science and Engineering, McGraw-Hill, New York, 1991, p. 335.

2. Wick, C., Benedict, J. T., and Veilleux, R. F., eds. Tool and Manufacturing Engineers Handbook, 4th ed., Vol. II: Forming, Society of Manufacturing Engineers, 1984.

3. Boyer, H. E., and Gall, T. L., eds. Metals Handbook—Desk Edition, American Society for Metals, Metals Park, OH, 1985, pp. 24–27, 26–50.

4. Schey, J. A. Introduction to Manufacturing Processes, 2nd ed., McGraw-Hill, New York, 1987, pp. 183–279.

5. Schey, J. A. Introduction to Manufacturing Processes, McGraw-Hill, New York, 1976, pp. 148–153.

6. Kalpakjian, S. Manufacturing Processes for Engineering Materials, 2nd ed., 1991, pp. 313–401.

7. Alexander, J. M., Brewer, R. C., and Rowe, G. W., Manufacturing Technology, Vol. 2: Engineering Processes, Wiley, New York, 1987, pp. 149–162.

8. Edwards, L., and Endean, M. Manufacturing with Materials, Butterworths, London, 1990, p. 174.

9. Knight, W. A., and Poli, C. "A Systematic Approach to Forging Design," Machine Design, Jan. 24, 1985, pp. 94–99.

10. Ludema, K. C., Caddell, R. M., and Atkins, A. G., Manufacturing Engineering, Prentice Hall, Englewood Cliffs, NJ, 1987, pp. 361–366.

11. Mielnik, E. M. Metalworking Science and Engineering, McGraw-Hill, New York, 1991, pp. 494–496.

12. Metals Handbook, Volume 1: Properties and Selection of Metals, 8th ed., 1961, American Society for Metals, Metals Park, OH, pp. 46–7.

13. Boyer, H. E., and Gall, T. L., eds. Metals Handbook—Desk Edition, 1985, American Society for Metals, Metals Park, OH, pp. 24–27, 26–50.

14. Meilnik, E. M. Metalworking Science and Engineering, McGraw-Hill, New York, 1991, p. 547.

15. Kalpakjian, S. Manufacturing Engineering and Technology, Addison-Wesley, Reading, MA, 1989, p. 393.

16. Altan, T., and Boulger, F. W. "Flow Stress of Metals and Its Application in Metal Forming Analysis," Transactions of ASME, Nov. 1973, pp. 1009–1018.

INTERNET SOURCES

History and definitions; links to other forging sources: *http://madmax.me.berkeley. edu/~mas/forging/*

Forging design: *http://www.temroc.com/noframes/page5.htm*

General description of rolling, hot die forging, isothermal forging, etc.: *http://www. cemr.wvu.edu/~imse304/*

Aalborg University Process Database with Forming/Shaping: *http://www.iprod. auc.dk/procesdb/index.htm*

13
Sheet Metal Forming

13.1 INTRODUCTION

Sheetmetal processing includes the sheet metal forming operations such as shearing, bending, drawing, spinning, roll forming, stretch forming, and bulging to form a wide variety of products. The distinguishing characteristics of sheet metal products is that they have a large surface area and a low modulus (ratio of volume to surface area) as compared to products produced by casting, bulk deformation, or machining. In general, sheet metal processes have two-dimensional deformation, because the thickness change is generally small, but the elastic recovery can be significant. Since the operations are often performed on presses, sheet metal forming is sometimes referred to as *pressworking*.

There are three major classes of sheet metal operations.

1. *Shearing operations*: A process of cutting sheet metal into smaller pieces, usually to be reworked by another process. Some of the typical operations are blanking, piercing, and slitting.
2. *Bending operations*: A process in which a straight workpiece is plastically deformed into an angle. Some of the typical operations are V-bending and press-brake bending.
3. *Drawing operations*: A process in which sheet metal is pressed into a shaped die to form an open-ended, cylindrical shape such as a can. Some typical operations are deep drawing, stretch drawing, and reverse drawing.

Compression operations, such as coining and extruding, are more like bulk deformation operations, although at least one author (1) considers them in the sheet metal category.

13.2 DEFECTS IN SHEET METAL PRODUCTS

There are several different types of defects that may occur in sheet metal products. Some of the more common defects (2) follow:

1. *Orange peel*: The outside surface becomes grainy. This can often be corrected by using a material of smaller grain size.
2. *Stretch marks (Lüders lines)*: These lines on the surface are harmless, but they are objectionable in appearance. Correct rolling can reduce the occurrence of these defects. The defect seems to be caused by nitrogen and carbon interstitials in steels.
3. *Localized thinning (necking)*: This occurs when the stress exceeds the ultimate strength or when surface defects permit a local stress concentration to occur.
4. *Earing*: When drawing a disk into a cup shape, the edges are not the same length and the edge gives an earlike appearance. This results from the planar anisotropy of the material, which causes the material to have different tensile strength values in different directions; the low-strength direction will elongate more than the high-strength direction. The planar anisotropy is often the result of improper rolling and annealing.
5. *Fracture*: This catastrophic defect results in the separation of the surface into one or more surfaces. Localized thinning leads to crack development, which leads to fracture.

These are the common defects that occur, and there are specialized defects, such as burrs (which occur during shearing) and wrinkling (which occurs in deep drawing). Fracture is the worst defect and is to be avoided, for the part is completely unusable and cannot be reworked. In some instances, a certain amount of thinning or earing may be permissible, and the surface defects of orange peel and stretch marks may be permitted on hidden surfaces.

13.3 STRAIN RATIOS AND ANISOTROPIC BEHAVIOR

The concepts of strain ratio, mean anisotropic coefficient, and planar anisotropic coefficient were previously presented, and some of the material will be repeated. The strain ratio and anisotropic coefficients have been used by the automotive industry to evaluate materials and processes for forming various sheet products, such as doors, fenders, roofs, hoods, and trunk lids. Extensive evaluations have been performed on low-carbon steels and aluminum alloys to predict formability characteristics.

In sheet metal products, the mechanical properties can vary directionally. The reference direction is generally the direction of rolling, and the
other commonly used directions are 45° and 90° from the rolling direction.
The expressions for true strain in the width and thickness directions and the
strain ratio R are:

$$\varepsilon_w = \ln(w_f/w_o) \tag{13.1}$$

$$\varepsilon_t = \ln(t_f/t_o) \tag{13.2}$$

$$R = \varepsilon_w/\varepsilon_t \tag{13.3}$$

where

R = strain ratio
ε_w = width strain
ε_t = thickness strain
w_o = original width before strain
w_f = final width after strain
t_o = original thickness before strain
t_f = final thickness after strain

The R value can be calculated for the various directions, and are designated
as:

R_0 strain ratio in rolling direction
R_{45} strain ratio at 45° to rolling direction on the rolling surface plane
R_{90} strain ratio at 90° to rolling direction on the rolling surface plane

These strain ratios can now be used to determine various anisotropic
coefficients. The two most commonly used coefficients are:

$$R_m = (R_0 + 2R_{45} + R_{90})/4$$

where

R_m = mean anisotropic coefficient

and

$$R_p = (R_0 - 2R_{45} + R_{90})/2$$

where

R_p = planar anisotropic coefficient

If the values of the strain ratio in the three directions are all equal to
1.0, the material is said to be *isotropic*. This would also cause the mean
anisotropic coefficient to have a value of 1.0, and the planar anisotropic
coefficient would be zero. If the values of the strain ratio in the three direc-

tions are all equal to a value other than 1.0, the material is said to have
planar isotropy; the planar anisotropic coefficient will be zero, but the mean
anisotropic coefficient will be a value other than 1 but equal to the individual
strain ratio value.

It is desirable to have a high value for the mean anisotropic coefficient,
for it implies a good resistance to thinning and a good thickness strength.
It is desirable to have a zero value for the planar anisotropic coefficient
because the larger the value, the greater the tendency for earing during deep

Table 13.1 Material Properties for Sheet Metal Materials

A. Strain ratios and anisotropic coefficients for typical engineering materials[a]

Material and structure		R_0	R_{45}	R_{90}	R_m	R_p
Normalized steel	BCC	0.9	1.1	0.9	1.0	−0.20
Copper, brass	FCC				0.5–0.9	
Zinc	HCP				0.2	
Lead	FCC				0.2	
Titanium	HCP				3.0–6.0	
Killed steel (A1)	BCC	1.6	1.4	1.9	1.6	+0.35
Rimmed steel	BCC	1.3	1.0	1.4	1.2	+0.35
Low-N, -C steels	BCC				1.8–2.4	
HSLA—hot rolled	BCC				0.8–1.0	
HSLA—cold rolled	BCC				1.0–1.4	

B. General sheet metal material properties[b]

Material	Tensile strain (UTS) e_u	Reduction of Area[c] (%) R_a	Ultimate tensile strength MPa	kpsi
Steel for drawing:				
Commercial quality	0.22	65	330	48
Drawing quality	0.24	75	310	45
Stainless steel (304)	0.40	75	515	75
Aluminum 1100	0.32	60	90	13
Copper	0.45	75	235	34
Titanium	0.20	40	345	50

[a]Adopted from Refs. 2, 3, and 8.
[b]Adopted from Ref. 10.
[c]Estimated values.

drawing. Typical strain ratio values and/or anisotropic coefficients for typical materials are presented in Table 13.1. Note that the material structure also has an influence upon the strain ratio and the anisotropic values. In general, FCC materials have lower values for the mean anisotropic coefficient than do the BCC materials. Hexagonal close-packed (HCP) materials with a high c/a ratio, such as zinc, have low values of the mean anisotropic coefficient, whereas HCP materials with a low c/a ratio, such as titanium, have the highest mean anisotropic coefficients. The final steps of the sheet rolling operation, specifically the tempering (heat treatment) conditions and the addition of strain to the material, are used to control the anisotropic coefficients.

In Fig. 13.1, the limiting draw ratio for a deep drawing operation is shown as a function of the mean anisotropic coefficient. The *limiting draw ratio* is the ratio of the maximum diameter of the blank to the punch diameter being used for the drawing operation; that is:

$$LDR = D_{(blank)}/D_p \qquad\qquad (13.4)$$

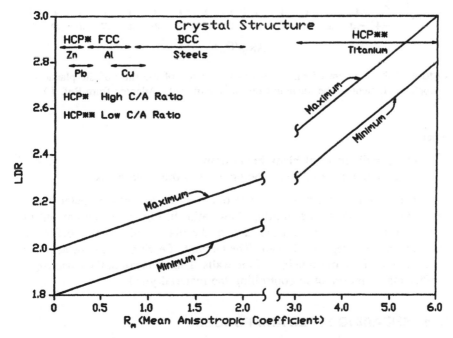

Figure 13.1 Variation in the limiting draw ratio (LDR) as a function of the mean anisotropic coefficient and the crystal structure. (From Refs. 8 and 9.)

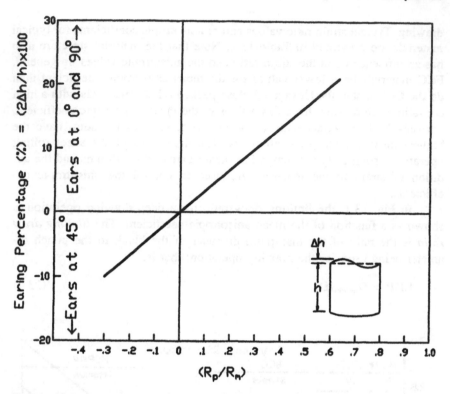

Figure 13.2 Amount of earing (%) as a function of the ratio of the planar anisotropic coefficient to the mean anisotropic coefficient. (Adapted from Ref. 4.)

where

$D_{(\text{blank})}$ = diameter of blank being drawn
D_p = diameter of punch being used to draw the blank

The crystal structure of the material is an important parameter in determining the value of the limiting draw ratio. In Fig. 13.2, the amount of
earing can be estimated from the ratio of the planar anisotropic coefficient
to the mean anisotropic coefficient. The amount of earing (4) can be as much
as 25 percent of the mean height of the walls. Thus, control of the anisotropic
coefficients is important in controlling the material yield.

13.4 SHEARING TYPES AND FORCES

The cutting of sheet metal is frequently done with shears to make strips or
discs for further processing. When strips are produced, the type of shear is

either guillotine or alligator. The guillotine shear had a notorious reputation for shearing items other than metals, and the alligator shear somewhat resembles a pair of scissors. The forces needed for shearing can be calculated by:

$$P_s = \text{shear stress} \times \text{shear area} \qquad\qquad (13.5)$$

When the shear stress values are not available, the following expression can be used to determine the shear stress:

shear stress = 0.7 × UTS (ultimate tensile strength)

shear area = length being cut × thickness

The length being cut frequently is the perimeter of the part, which would be πD ($\pi \times$ diameter) for circular parts.

The difference between the two types of shear is the length being cut. In Fig. 13.3, the two areas are illustrated. The guillotine shear cuts the entire length, whereas the alligator shear is cutting only a portion at a time and

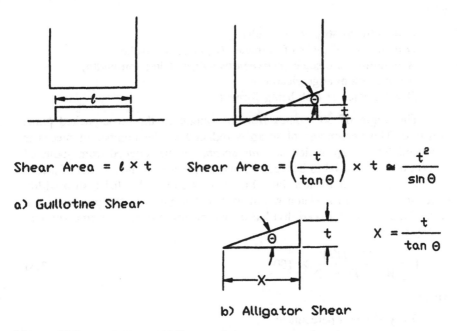

Figure 13.3 Guillotine and alligator shear areas.

requires lower forces. For the alligator shear, the length being cut can be expressed as:

$$x = t/\tan \Theta \tag{13.6}$$

where

    t = thickness of material
    Θ = angle of shear blade
    x = length being cut at any instant

If the angle of the shear blade changes, then the forces required for shearing will also change. A pair of scissors, for example, starts with a large angle; then the angle decreases during the cutting. Paper cutters generally have a curved blade, the purpose of which is to keep a constant angle during the cutting operation so the forces remain relatively constant during cutting (shearing).

In presses, several parts may be sheared at a time, and the more pieces, the greater the yield of the process. Formulas have been developed for the shearing of circular blanks from sheet metal strips. The width of the strip can be calculated (5) as:

$$W = (N - 1) \times \sin 60° \times (D + X) + (D + 2 \times Y) \tag{13.7}$$

where

    W = width of strip (coil width)
    N = number of disks of diameter D per press stroke
    X = minimum distance between two disks (skeleton width)
    Y = minimum edge clearance
    D = diameter of disk being formed

The angle of 60° comes from the angle in the close-packed plane structure. The percentage of scrap is reduced as the number of pieces is produced. The scrap, called *skeleton scrap*, can be reduced from about 24 percent of the material to between 12 and 13 percent as the number of pieces increases from 1 to 10 at a time. Figure 13.4 shows the disks, strip width, edge clearance, and skeleton width values. The yield of the process is the area of the disks produced divided by the corresponding coil area and can be expressed as:

$$Y = \frac{N \times \pi D^2/4}{W \times (D + X)} \times 100 \tag{13.8}$$

where

    Y = yield as a percentage
    N = number of disks per repetitive stroke of press

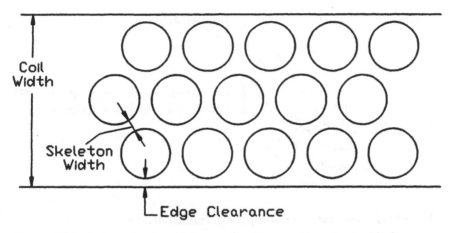

Figure 13.4 Layout for three-disk operation for punching circular blanks.

D = diameter of disk being formed
X = minimum distance between two disks (skeleton width)
W = width of strip (coil width)

As the number of disks per stroke increases, the process yield increases but the amount of increase decreases. In addition, the size of the press needed would increase; thus, there is an "optimal" width of coil to be used and corresponding number of disks produced per press stroke.

13.5 BENDING STRESSES, MINIMUM BEND RADIUS, AND BEND LENGTH

Bending is probably the most common sheet metal operation for forming shapes, with the exception of shearing. Two of the factors in evaluating bending are the minimum radius of the bend angle to prevent cracking and the determination of the required strip length prior to bending to produce the final shape, which is called the *bend length*. The minimum bend radius is also a function of the thickness of the material, so the ratio of minimum bend radius to metal thickness, R_b/t, is evaluated. There are two methods used to evaluate the minimum bend radius, one based upon the engineering strain at the ultimate tensile strength, and the second based upon the percent reduction of area and the true strain.

The minimum bend radius based upon the engineering strain limits the strain in the outer fibers of the bend to the strain at the ultimate tensile stress, which is where necking or localized thinning starts. The parameters indicated in Fig. 13.5 are the material thickness (t), the radius of the bend

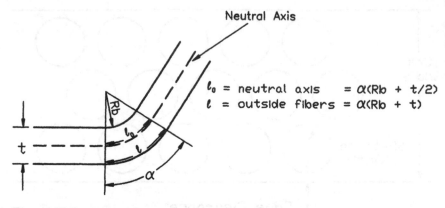

Figure 13.5 Bend radius, bend angle, and material thickness.

(R_b), and the bend angle (α). The bend radius ratio expression can be obtained from the engineering strain at the UTS via:

$$e_u = (l - l_o)/l_o \qquad (13.9)$$

$$e_u = \frac{\alpha(R_b + t) - \alpha(R_b + t/2)}{\alpha(R_b + t/2)}$$

$$e_u = \frac{t/2}{R_b + t/2}$$

$$e_u = \frac{1}{2R_b/t + 1} \qquad (13.10)$$

where

    e_u = engineering strain at UTS
    l = length of outside fiber after bend
    l_o = length of neutral axis (original length of outside fiber)
    α = angle of bend
    t = thickness of material being bent
    R_b = radius of bend, or bend radius

It is important to note that Eq. (13.10) is independent of the bend angle (α). Equation (13.10) can be rewritten in terms of the bend radius ratio as:

$$R_b/t = 1/2 \times (1/e_u - 1) \qquad (13.11)$$

This indicates that for brittle materials, where the engineering strain at the UTS would be low, a high bend radius ratio would be required, whereas ductile materials with a large engineering strain would not require a high bend radius ratio. Some estimates of engineering strains at the UTS are given in Table 13.1.

The second approach is based upon the true strain and the reduction of area. The true strain can be expressed in terms of the original and final fiber lengths and in terms of areas as:

$$\varepsilon = \ln(l/l_o) \tag{13.12}$$

$$\varepsilon = \ln\left(\frac{R_b + t}{R_b + t/2}\right) \tag{13.13}$$

$$\varepsilon = \ln(A_o/A) \tag{13.14}$$

where

l = length of outside fiber after bend
l_o = length of neutral axis
R_b = radius of bend
t = material thickness
A = cross-sectional area after bend
A_o = cross-sectional area before bend

The percentage reduction of area, R_a, can be expressed in terms of the cross-sectional areas as:

$$R_a = (A_o - A)/A_o \times 100 \tag{13.15}$$

This permits the determination of the ratio A_o/A in terms of R_a as:

$$A_o/A = 100/(100 - R_a) \tag{13.16}$$

If one sets equal to each other the expressions for A_o/A in Eqs. (13.13) and (13.16), the relationship between the reduction of area and the bend radius ratio can be obtained; that is, if:

$$100/(100 - R_a) = \frac{R_b + t}{R_b + t/2}$$

then one can obtain:

$$R_b/t = 50/R_a - 1 \tag{13.17}$$

where

R_b/t = bend radius ratio
R_a = reduction of area as percentage

This expression and typical values of the bend ratio for different materials are presented by Kalpakjian (4). Experimental data (4) has been obtained that tends to validate the reduction of area expression, except that a value of 60 is used instead of 50 in Eq. (13.17). Values of the reduction of area, R_a, are included in Table 13.1.

The bend length expression is used to determine the linear length needed for a given bend radius, angle of bend, and material thickness. The bend length can be calculated from:

$$L = \alpha/360 \times 2\pi \times (R_b + kt) \tag{13.18}$$

where

L = length of material for bend
α = angle of bend
R_b = radius of bend
t = material thickness
k = 0.33 for tight bends where $R_b < 2t$, or 0.50 for regular bends where $R_b \geq 2t$

In Fig. 13.6, various bend angles are illustrated; at the second bend, the angle is greater than 90°. One of the common errors is to take the internal angle at the second bend rather than at the actual bend angle. The bend angle typically has an upper limit of 180°. The value of k is 0.50 for regular bends, since the neutral axis is at the middle of the material. For tight bends, the material cannot move as easily and the net effect is a movement of the neutral axis toward the inside bend surface, and the value of k is 0.33 for tight bends.

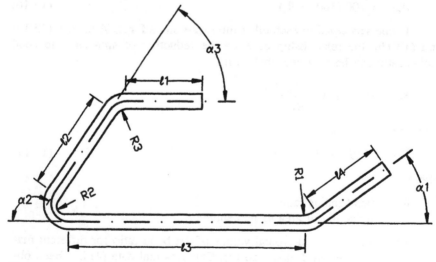

Figure 13.6 Bend length calculation example.

Example Problem 13.1. If the material thickness is 3 mm and the linear lengths, bend radii, and bend angles in Fig. 13.6 have the following values, what is the total length?

Values for Bend Length Calculations for Example Problem 13.1

Linear lengths	Bend radii	Bend angles (degrees)
l_1 = 20 mm	R_1 = 10 mm	α_1 = 40°
l_2 = 40 mm	R_2 = 5 mm	α_2 = 130°
l_3 = 80 mm	R_3 = 8 mm	α_3 = 50°
l_4 = 50 mm		

$$
\begin{aligned}
L \text{ (mm)} &= (20 + 40 + 80 + 50) + (40/360)2\pi(10 + 0.50 \times 3) \\
&\quad + (130/360)2\pi(5 + 0.33 \times 3) + (50/360)2\pi(8 + 0.50 \times 3) \\
&= 190 + 8.02 + 13.61 + 8.29 \\
&= 190 + 29.92 \\
&= 219.92 \\
&= 220 \text{ mm}
\end{aligned}
$$

13.6 DEEP DRAWING CALCULATIONS

Deep drawing is the process in which a blank is drawn by a punch into a hollow die to form a container with walls. Most of the shapes are circular, such as cans, cooking pots, and cartridge shells, but rectangular shapes such as baking containers or pans can also be deep drawn. An illustration of the deep drawing equipment and terms can be seen in Fig. 13.7. The clearance is the difference between the die and punch radii, or the die and punch diameter difference divided by 2.

The formulas to calculate the initial blank diameter for various designs are presented in Fig. 13.8. These expressions assume that the thickness does not change, and expressions for other designs are presented in Refs. 4 and 5. These expressions can be derived from the assumption that the surface areas before and after drawing are equal. Once the blank diameter has been determined, the forces needed to shear the blank can be calculated using Eq. (13.5), where the length of cut is the circumference of the blank. The thickness of the bottom may be different than the wall thickness, as, for example, in the production of cooking pots. For these instances, the volumes, not the areas, are assumed to be constant before and after drawing. The thickness of the walls can be less than the thickness of the base, due to ironing that

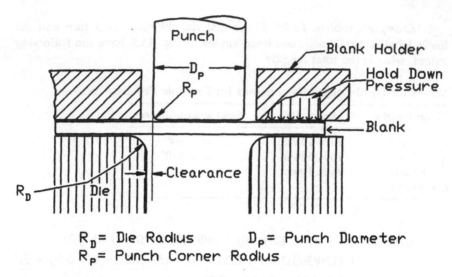

R_D = Die Radius D_P = Punch Diameter
R_P = Punch Corner Radius

Figure 13.7 Schematic illustration of deep drawing operation.

can occur during the drawing operation. The uniform ironing will occur because the clearance is less than the thickness of the blank being drawn. For a simple shape with no corner radii, the expressions would be:

$$V(\text{before}) = V(\text{after})$$

which results in:

$$\pi D^2_{(\text{blank})}/4 \times t_{(\text{base})} = (\pi D^2_a/4) \times t_{(\text{base})} + \pi D_a H \times t_{(\text{wall})}$$

which can be reduced to:

$$D_{(\text{blank})} = \sqrt{D^2_a + 4D_a H \times t_{(\text{wall})}/t_{(\text{base})}} \qquad (13.19)$$

where

$D_{(\text{blank})}$ = diameter of blank
D_a = diameter of punch (or part) after drawing
H = height of wall of cup after drawing
$t_{(\text{base})}$ = thickness of blank or base of part
$t_{(\text{wall})}$ = thickness of wall after drawing

Example Problem 13.2. A small cup with a diameter of 50 mm and a wall height of 75 mm is to be deep drawn from an aluminum sheet that is 2 mm thick. What should the initial diameter of the blank be?

Shape   Formula for Blank Diameter [D$_{(Blank)}$]

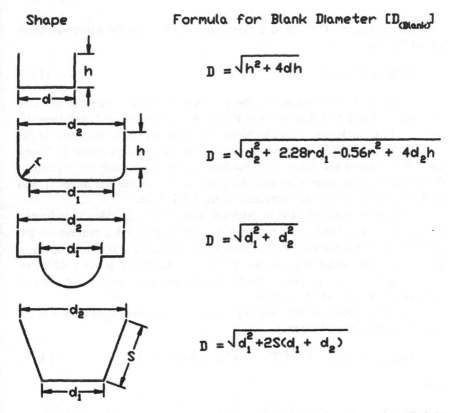

$$D = \sqrt{h^2 + 4dh}$$

$$D = \sqrt{d_2^2 + 2.28rd_1 - 0.56r^2 + 4d_2h}$$

$$D = \sqrt{d_1^2 + d_2^2}$$

$$D = \sqrt{d_1^2 + 2S(d_1 + d_2)}$$

Figure 13.8   Shapes and blank diameter formulas for deep drawing. (From Ref. 5.)

Using Eq. (13.19),

$$D_{(blank)} = \sqrt{50^2 + (4)(50)(75)} \times 2/2$$
$$= 132.3 \text{ mm}$$

The draw ratio, which is the diameter of the blank disk to the diameter of the punch, is controlled by the limiting draw ratio. The limiting draw ratio is determined by the mean anisotropic coefficient of the material as indicated by Fig. 13.1 and is between 2 and 3 for nearly all metals. If the draw ratio is greater than the limiting draw ratio, this implies that the drawing of the part will require more than one drawing operation. That is, redrawing will be necessary. Long, thin, tubular-shaped products, such as cartridge shells and metal mechanical pencils or pens, require more than one drawing operation. The limiting draw ratio (LDR) can be expressed from Eq. (13.4) as:

$$\text{LDR} = D_{(\text{blank})}/D_p \tag{13.4}$$

If the LDR is exceeded for the material, the punch diameter must be increased for the first draw so that the LDR is not exceeded, and then the part would be redrawn, with the blank diameter being the diameter of the blank after drawing. The drawing ratio for redrawing operations is lower, since the material has been strain hardened to approximately 50 percent of the previous draw. Annealing may be performed to soften the material, and then the draw ratio can be increased to the LDR value.

The blank must be held in position with the blankholder as indicated in Fig. 13.7. The hold-down pressure is approximately 1.5 percent of the yield stress of the material being formed. If the blank is not held down, a defect known as *wrinkling* will occur. On the other hand, if the hold-down pressure is too great, the blank will fracture in the corner or even be sheared out and the press will be jammed.

The maximum drawing force for the deep drawing operation can be estimated using Eq. (13.20):

$$F(\text{max}) = \pi D_p \times t \times \text{UTS} \times [(D_{(\text{blank})}/D_p) - 0.7] \tag{13.20}$$

where

$F(\text{max})$ = drawing force (maximum)
D_p = diameter of punch
t = thickness of blank
UTS = ultimate tensile strength of material
$D_{(\text{blank})}$ = diameter of blank being drawn

This expression assumes that good lubrication exists between the blankholder and the blank and between the blank and the die. The drawing forces will often be greater than the shearing forces; but the shear has a sharp cutting edge, whereas the punching die has rounded corners. If lubrication is not present, the punching die can rupture the part in the corner, and not only is the part ruined, but the press is also jammed with the ruptured part. Some typical UTS values for drawing materials are included in Table 13.1.

In the drawing operation, there frequently is reference to the reduction of the operation. The percent reduction is usually based upon diameters (linear reduction) for circular parts and based upon area for rectangular parts. The two different types of reduction can be illustrated by the following expressions:

Diameter-based reduction:

$$R(\%) = (D_{(blank)} - D_p)/D_{(blank)} \times 100 \qquad (13.21)$$

where

$R(\%)$ = percent reduction
$D_{(blank)}$ = diameter of blank
D_p = punch diameter

Area-based reduction:

$$RA(\%) = [(A_{(blank)} - A_{(base)})/A_{(blank)}] \times 100 \qquad (13.22)$$

where

$RA(\%)$ = percent reduction of area
$A_{(blank)}$ = area of blank being drawn
$A_{(base)}$ = bottom area (base of drawn part

Figure 13.9 illustrates the different areas for rectangular and circular parts. In rectangular parts, there is excess material in the corners, and this material must be removed. Sometimes material is removed from the blanks prior to drawing to reduce the excess material in the corners. The excess material increases as the wall height of the rectangular parts increases. With circular parts, the amounts of excess material are considered to be very small.

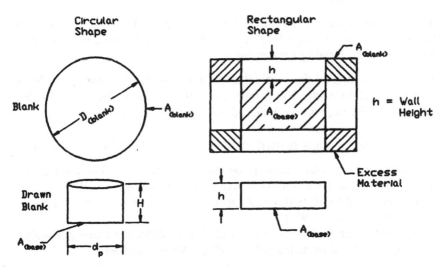

Figure 13.9 Area values used for reduction of area calculations and excess material in rectangular shapes.

13.7   SUMMARY AND CONCLUSIONS

Sheet metal operations are the most common methods for producing thin parts with a high surface-area-to-volume ratio (or low modulus). The basic operations of shearing, bending, and drawing were considered, and the basic calculations used with these processes were presented. Most of the operations require a large amount of tooling, so high production quantities are required to offset the tooling costs.

The forces required for an alligator shear are lower than those for the guillotine shear, but for many products the guillotine type must be used to prevent bending of the part. The key in bending operations is to prevent rupture of the outside fibers. Deep drawing is the primary method for producing walled containers such as cans, pots, and cartridge shells. It requires an optimal range for forces: sufficient force to form the shape and limited force to prevent rupture of the part. Deep drawing is generally restricted to metals, because other materials do not possess the proper combination of ductility and stiffness to be formed in this manner.

13.8   EVALUATIVE QUESTIONS

1. What are the three classes of sheet metal working operations?

2. What are the general types of defects that occur during sheet metal working operations?

3. If a material has the strain ratio values of $R_0 = 1.43$, $R_{45} = 1.20$, and $R_{90} = 1.50$, determine the values of R_m and R_p, and estimate the LDR value and the amount of earing (percentage).

4. For the listed conditions, calculate the force needed to shear a strip from a coil that is 12 in. wide, 100 ft long, and 0.030 in. thick. The strip is to be $2 \times 12 \times 0.030$ in., and the material has a UTS value of 50,000 psi.
   a. A guillotine shear is used.
   b. An alligator shear is used with an angle of 10°. (12,600 lb, 181 lb).

5. A blank 4 in. in diameter is to be sheared from a coil. The edge clearance is 0.250 in. and the minimum width between blanks is 0.300 in.
   a. What is the coil width and yield if one disk is made per width of coil? (4.5 in., 64.9%)
   b. What is the coil width and yield if three disks are made per width of coil at a time at the 60° angle? (11.94 in., 73.4%)

6. A material has an engineering strain value of 0.240 at the UTS value of 50,000 psi, and the angle of bend is 40°. What is the bend radius ratio?

7. If the reduction of area for a material with the UTS of 70,000 psi is 30 percent and the angle of bend is 25°, what is the bend radius ratio?

8. Redo Example Problem 13.1 with the bend angles being 50°, 150°, and 30° for the three angles and the material thickness of 2 mm. Assume all other values remain the same.

9. A sheet metal part has a thickness of 0.100 in., and the material has a UTS of 40,000 psi, an ultimate strain of 0.200 in./in., and a reduction of area of 20 percent at the UTS. Calculate the minimum bend radius based upon:
   a. Ultimate strain (0.20 in.)
   b. Reduction of area (0.15 in.)

10. A can is to be made with a wall height of 5.0 in. and a diameter of 2.0 in. The sheet metal is 0.015 in. thick, and the wall and base are to be the same thickness. The material has a yield strength of 30,000 psi and an ultimate tensile strength of 40,000 psi. The material is on a coil of strip form, which is 10 ft in length and 10 in. wide.
   a. What should be the starting blank diameter, in inches? (6.63)
   b. What force is needed to shear the blank from the strip with a guillotine-type shear punch? (8736 lb)
   c. What is the draw ratio for the can? (3.31)
   d. What is the percent reduction in diameter? (69.8%)
   e. What force is needed to draw the can in pounds? (9858 lb)
   f. If the draw ratio was limited to 2.2, what force would be needed for the first draw, in pounds? (8540)

11. A sheet 0.020 in. thick is to be used to make a can with a wall thickness of 0.010 in. and a bottom thickness of 0.020 in. If the bottom is 4.0 in. in diameter and the height is 10 inches, then:
   a. What is the diameter of the blank sheet needed to make the can? (9.8 in.)
   b. What is the draw ratio for making the can? (2.45)

12. Assume the values of Example Problem 13.1 are:
$l1 = 25$ mm	$R1 = 8$ mm	$\alpha1 = 30$	$t = 3$ mm
$l2 = 50$ mm	$R2 = 10$ mm	$\alpha2 = 120$	
$l3 = 15$ mm	$R3 = 10$ mm	$\alpha3 = 90$	
$l4 = 10$ mm			

   a. What is the total length of the strip needed to make the shape? (147 mm)
   b. What is the engineering strain and the true strain in the outer fiber at the bend, where $R = 8$ mm? (0.157, 0.147)

REFERENCES

1. Mielnik, E. M. Metalworking Science and Engineering, McGraw-Hill, New York, 1991, pp. 689–957.
2. Kalpakjian, S. Manufacturing Processes for Engineering Materials, 2nd ed., Addison-Wesley, Reading, MA, 1991, p. 453.
3. Keeler, S. P. "Understanding Sheet Metal Formability," Machinery, Series of Articles during February to July, 1968.
4. Ludema, K. C., Caddell, R. M., and Atkins, A. G. Manufacturing Engineering, Prentice Hall, Englewood Cliffs, NJ, 1987, pp. 361–366.
5. Tooling and Manufacturing Engineers Handbook, 4th ed., Volume 2: Forming, Publisher & City needed, p4:39–40.
6. Kalpakjian, S. Manufacturing Processes for Engineering Materials, 2nd ed., Addison-Wesley, Reading, MA, 1991, p. 419.
7. Green, R. G. "Using Ironworkers for One-Stop Chopping," Forming and fabricating, June/July 1994, pp. 16–19.
8. Atkinson, M. "Assessing Normal Anisotropic Plasticity of Sheet Metals," Sheet Metal Industries, Vol. 44, p. 167.
9. Schey, John A. Introduction to Manufacturing Processes, 2nd ed., McGraw-Hill, New York, 1987, pp. 280–329.
10. Boothroyd, G., Dewhurst, P., and Knight, W. Product Design for Manufacture and Assembly, Marcel Dekker, 1994, p. 363.

INTERNET SOURCES

General sheet metal design, helix modeling of sheet metal parts: *http://www.spi.de/sheetmetal/sm1.htm*
Bending: *http:www.amada.net/htm/rg/rgABCtoc.htm#top*
Deep drawing: *http://www.htw-dresden.de/~manufact/ote/hptze.htm*
Shearing: *http://www.iprod.auc.dk/procesdb/shearing/start.htm*
Grain size effects in sheet metal forming: *http://www.cdw.de/english/stiefzie.htm*

APPENDIX 13.A   ALTERNATIVE SHEARING
FORCE FORMULA

An expression presented by Green (7) for shearing sheet metal is similar to Eq. (13.5). It is, for the English system of units:

$$P_s(\text{tons}) = 80 \times \text{punch dia (in.)} \times \text{material thickness (in.)} \qquad (13.5a)$$

This expression is for mild steel with a UTS of 65,000 psi. For other materials, a multiplier is used:

Material	Multiplier	Material	Multiplier
Aluminum	0.38	Steel (mild)	1.00
Brass	0.70	Steel (0.5 C)	1.50
Copper	0.56	Stainless (303)	1.50
		Steel (cold drawn)	1.20

What would be the value of the constant (80) in Eq. (13.5a) if cm and mPa are used?

14

Machining (Material Removal or Cutting) Processes

14.1 INTRODUCTION

Machining involves the shaping of a part through removal of material. A tool, constructed of a material harder than the part being formed, is forced against the part, causing material to be cut from it. Machining, also referred to as *cutting, metal cutting,* or *material removal,* is the dominant manufacturing shaping process. It is both a primary as well as a secondary shaping process. *Machining* is the term generally used, rather than *material removal* or *cutting.* The device that does the cutting or material removal is known as the *machine tool.* Nearly all castings and products formed by deformation processing (bulk or sheet metal) require some machining to obtain the desired final shape or surface characteristics.

The difference between machining and the other two primary forming processes is that materials are removed from a general bulk shape to obtain the desired three-dimensional shape. The material is generally removed in the form of "chips," but it may also be removed as fine particles or powder (in grinding or polishing operations), or by dissolving or vaporizing (in processes such as electrochemical machining and electrical discharge machining). The removed material, chips, powders, etc., is difficult to recycle and can become expensive waste materials because of environmental problems. Shearing is a cutting operation that is generally classified as a sheet metal forming operation, because sheet metal parts must be cut out before they are formed, but it is actually a machining operation.

The primary reasons for selecting machining over the other two primary shaping processes are that machining can:

1. Improve dimensional tolerances
2. Improve surface finish

3. Produce complex geometry such as:
    a. Holes (generally circular, but also other shapes)
    b. Sharp corners and flat surfaces
4. Produce low quantities economically because of:
    a. More flexibility in tooling and fixturing
    b. Low operating costs (equipment)
5. Lower setup times (time to prepare tooling for production)

In many cases, machining is a secondary operation for casting and forming processes, to obtain the required dimensional tolerances, surface finish, or complex geometry of the part.

Machining processes also have problems that must be considered.

1. Large amounts of chips, etc., which cause:
    a. Poor material utilization (over 60 percent of the material becomes chips in typical light machining operations).
    b. Waste scrap/material disposal problems (cutting fluids must be environmentally safe).
2. High energy consumption
    a. Large amounts of chip surface are generated.
3. Longer cycle times
    a. Processing times are generally long, because several cuts are needed.
    b. Several operations are performed rather than one major operation.

Machining is the only primary forming process that is also used for secondary operations. This unique characteristic has led to the dominance of this process. Due to the high cost of machining and problems caused by the chips produced, casting and deformation processing try to produce "near-net shape" products, which can be completed with little or no machining.

14.2 CHIP FORMATION

There are three main variables that affect the formation of the chips in cutting:

1. Properties of the work material
2. Properties of the tool material
3. Tool geometry

The interaction between the tool and work material is also significant; this is often mentioned as the fourth main variable.

The tool geometry is described by the various angles and nose radius of the single-point tool illustrated in Fig. 14.1. The tool wear occurs on the flank face (flank wear) or on the top face (crater wear). The tool signature consists of the six angles that define the tool surface and the nose radius. Insert tools often have several of the angles at zero degrees.

The types of chips formed are generally classified into three types (1):

1. *Discontinuous (segmented) chips*: from hard, brittle materials and from two-phase materials that separate easily, such as leaded steels and gray cast irons.
2. *Continuous chips*: sharp, long, continuous chips, which can be sharp and hot and thus dangerous, steel and aluminum.
3. *Built-up edge*: part of the chip adheres to the tool, which produces rough surfaces on the finished part.

Discontinuous chips are preferred, because they tend to leave good surfaces and because of the safety aspect. Chip breakers are used to prevent the continuous chips from becoming dangerous. Elements are often added to steel to make it more machinable. Some of the common elements added are lead and sulfur.

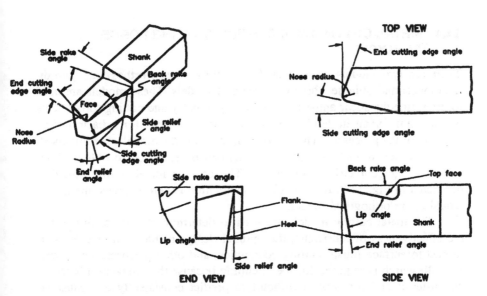

Figure 14.1 Tool signature: angles and nose radius.

14.3 MACHINABILITY

Machinability is an index of the relative ease with which a chip can be separated from the base material. The higher the machinability, the easier it is to machine the material. The reference steel B1112 is an easily machined material and is given a value of 100. Good machinability values range between 100 and 130, and for poor machinability the values range from 30 to 50.

There are various characteristics used to measure machinability, and no one characteristic has been adopted as the standard. Some of these (2) follow.

- a. Cutting speed for a 60-min tool life
- b. Volume of material removed in the tool life
- c. Tool life, in minutes
- d. Tool forces, energy, or power required
- e. Temperature of metal

Thus, when comparing metals with different machinability values, one should know which characteristic was used to determine the machinability value. Material ranking by one characteristic may not be the same as by another characteristic.

14.4 METAL-CUTTING MODELS FOR CUTTING FORCE ANALYSIS

There are two models used to analyze cutting forces: orthogonal (two-dimensional) and oblique (three-dimensional) models. These models are used to determine the shear angle from the tool geometry and cutting forces. This is important for an in-depth analysis of cutting but not for a basic understanding of the processes. The orthogonal model is the model generally used, for it is easier to analyze, but the results can be extended to oblique (3D) cutting using geometric relationships. The oblique metal-cutting model is illustrated in Fig. 14.2; if the angle of inclination is zero degrees, the model would be an orthogonal model.

The metal-cutting models are used to determine the forces in cutting, the effectiveness of lubricants, the optimal design of the cutting edge with respect to surface finish, cutting economics, and other performance parameters. Research is required for metal cutting because the interaction between the material and the tool is difficult to predict consistently and must be documented experimentally.

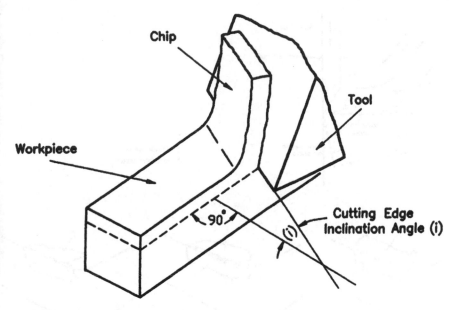

Figure 14.2 Oblique cutting at inclination angle (*i*).

14.5 TOOL WEAR FAILURE MECHANISMS

The tools can fail via various mechanisms. There are five different mechanisms, classified into two categories: primary and secondary. The primary failure mechanisms are:

1. Flank wear (clearance face wear, abrasion)
    a. Rough cuts: 0.030 in. (0.76 mm)
    b. Finish cuts: 0.015 in. (0.38 mm)
2. Crater wear (adhesion)

The secondary failure mechanisms are:

3. Oxidation
4. Breakage (shock, fatigue)
5. Chipping of tool (chatter, vibrations, etc.)

Flank wear is the type of wear that is generally implied when wear is being discussed; however, crater wear (3) is frequently the controlling mode when very high-speed cutting conditions are used. These two types of primary wear are illustrated in Fig. 14.3.

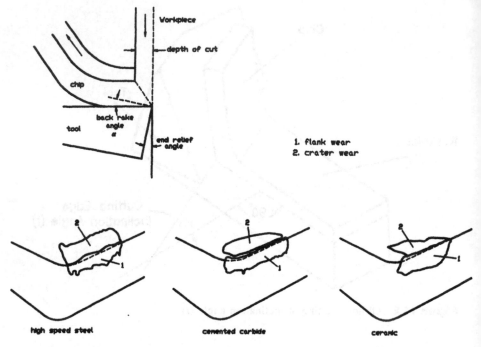

Figure 14.3   Types of wear observed in single-point cutting tools.

14.6 TAYLOR TOOL-LIFE MODEL

The Taylor tool-life model generally deals with flank wear. Crater wear models have also been formulated using the Taylor tool-life model. The basic expression for the Taylor tool-life equation is:

$$VT^n = C \tag{14.1}$$

where

V = cutting speed (ft/min or m/sec)
T = tool life (min or sec)
n = Taylor tool-life exponent
C = Taylor tool-life constant (ft/min or m/sec)

Many students, particularly those who have had courses in dimensional analysis, have extreme difficulty with this relationship, because the units are not consistent on both sides of the equation. This is not a theoretically derived equation, but a relationship derived from the observation of experimental data (by Frederick W. Taylor in the late 1800s). The units of V and C must be the same, and the units of time must be the same as those for V

and C. For those who cannot accept such inconsistency, the following relationship is presented:

$$V(T/T_o)^n = C \qquad (14.1a)$$

where the variables are the same as for Eq. (14.1) and

T_o = reference time unit

= 1 min or 1 sec, the same time unit as for T, C, and V

U.S.-English units of velocity are usually ft/min, and the time units are minutes. The ISO-metric units of velocity are generally m/sec, and the tool life time is in seconds. The generalized Taylor tool-life equation includes variations in feed and depth of cut as well as in cutting speed:

$$TV^{1/n1}f^{1/n2}d^{1/n3} = C \qquad (14.2)$$

where, in general,

$$n1 < n2 < n3 < 1 \quad \text{and} \quad 1/n1 > 1/n2 > 1/n3 > 1$$

so cutting speed (V) is the most important cutting variable, feed (f) is the next most important cutting variable, and depth of cut (d) is the least important of the three cutting variables with respect to tool life.

Example Problem 14.1. If the values of C and n for the basic Taylor tool-life equation are 200 ft/min and 0.25, respectively, answer the following questions.

  a. A cutting speed of 50 ft/min is used. What is the tool life?
  b. The cutting speed is increased 20 percent (to 60 ft/min). What is the effect upon tool life?

Solution.

  a. If $VT^{0.25} = 200$, then:

$$T = (200/V)^{1/n} = (200/50)^{1/0.25} = 4^4$$
$$= 256 \text{ min}$$

  b. If $V_1 T_1^n = V_2 T_2^n = C$, then:

$$T_2/T_1 = (V_1/V_2)^{1/n_1} = (V_1/1.2V_1)^4 = 0.48$$

That is, the tool life is only 48 percent of the original value; a 20 percent increase in cutting speed has resulted in a 52 percent decrease in tool life. For the specific tool life in part (a), the tool life is:

$$T_2 = 256 \times (50/60)^{1/0.25} = 123 \text{ min}$$

Increases in cutting speed cause large decreases in tool life, especially when the Taylor tool-life exponent is small. The effect of changes in cutting speed upon tool life is much greater for tool steels, which have low values of n, than for carbides or ceramics, which have higher values of n.

14.7 CLASSIFICATION OF MACHINING PROCESSES

There are several different classifications of machining processes. One classification is by the type of cutting tools; a second is by the type of surface generated. The classification by the type of surface generated is more important, because the surface of the product is one of the major criteria considered in the selection of the manufacturing process. The classification based on the type of cutting tool considers the number of cutting edges of the tool. The types are:

1. *Single-point cutting*: processes such as turning, planing, shaping, and boring
2. *Multiple-point cutting*
   a. Two edges: drilling
   b. n edges: milling, sawing, reaming, broaching, etc.
   c. Infinite number of edges: grinding, polishing

The surfaces generated have been classified into four types. The machining processes that generate the surface types are presented in Table 14.1. The types of surfaces generated are somewhat analogous to the concept of features, and these may become the type of features used to describe part surfaces in manufacturing.

The third approach to classifying processes is to list the processes and describe their capabilities. This is the most common approach to classifying processes, but it is difficult to evaluate which process is best for specific operations. The basic metal-cutting processes of turning, drilling, shaping and milling will be evaluated by this procedure.

1. *Turning*:   The turning operation is performed on a lathe, where the workpiece rotates and the tool moves parallel to the center axis of the workpiece. The operation is used to produce external cylindrical surfaces for parts such as shafts and axles. The key parameters of the lathe are the *swing*, which indicates the maximum diameter that can be turned, and the *length between centers*, which indicates the maximum length that can be turned. The surface finish of the turning operation is between 500 and 1000 μin. (12.5–25 μm) for rough turning and between 32 and 125 μin. (0.81–3.2 μm) for finishing operations. In addition to turning, operations such as drilling and facing can be done on a lathe, but the primary operation is turning.

Table 14.1 Surface Classification and Processes Capable of Generating Specific Surfaces

Surface type	Description of generated surface	Processes capable of generating surface type
1	Flat surface	
	a. Multiple (2 or 3)	Shaping, planing, milling, sawing, grinding
	b. Face (1 side)	Same as for (a) plus turning
2	Cylindrical surface	
	a. Internal surface	Drilling (via lathe, mill, and drill press, reaming, boring, tapping
	b. External surface	Turning
3	Surfaces of revolution (cones, cams, spheres)	Turning (with taper gage), milling
4	Wavy, irregular surfaces	Milling, grinding, sawing (2D)

2. *Drilling*: The purpose of the drilling operation is to create holes, and the instrument generally used is the drill press. There are two main types of equipment used for the drilling operation: the bench drill press and the radial drill press. The bench drill press is used for small parts, where the size of the drill press is twice the distance from the spindle to the column. The size is typically 10–20 in. (25–50 cm) for most drill presses, whereas the size of the radial drill press is determined by the length of the arm, which can be more than 72 in. (or 2 m). The surface finish from drilling is typically between 63 and 250 μin. Improved surface finish can be obtained by reaming, which is an operation used to finish holes, whereas drilling is used to create the hole.

3. *Shaping*: The shaper is a relatively simple tool. The workpiece is held in a vice while the ram, which carries the tool, slides back and forth in equal strokes to the desired length of stroke. The tool cuts in only one direction, but the return stroke is faster than the cutting stroke, to reduce idle time. This is known as the return:cutting-speed ratio. The shaper has two different types of drives, mechanical and hydraulic. The mechanical shaper is cheaper and uses mechanical drives to move the ram back and forth. The hydraulic drive has a quicker return speed than the mechanical drives, and the cutting speed and pressure are constant in the ram, but it is significantly more expensive. Shapers are used mainly for facing, but they can also be used for slots, to create steps, and for dovetails. There are two

kinds of shapers, vertical (slotter) and horizontal. The two machines are similar except for the motion of the ram.

4. *Milling*: Milling is an operation using a multiple-tooth cutter (tool), predominately to generate flat surfaces, but also to generate complex surfaces. The milling machine is classified as either a horizontal or a vertical machine. Horizontal milling has the axis of rotation of the cutter parallel to the milling table; a vertical milling machine has the axis of rotation perpendicular to the milling table. The milling table (or bed) is approximately 1 ft wide and 4 ft long (25 cm × 100 cm), but it can be much larger for special applications. When a horizontal machine is used, there are two types of cutting:

1. *Conventional (up) milling*: The chip formed starts thin and then increases as the cutter advances into the work.
2. *Climb (down) milling*: The chip formed starts thick and decreases as the cutter advances to the finished surface.

The cutters used on horizontal milling machines are generally plain cutters (also called *slab mills*) or side cutters. Generally, flat surfaces or slots are produced, but form cutters can be used to generate contoured surfaces.

A vertical milling machine uses end mills and face cutters. It frequently is used to finish surfaces or to produce slots and dovetails. The vertical milling machine can also be used for drilling operations when the table has vertical feed. Figure 14.4 illustrates vertical and horizontal milling and the two types of cutting used in horizontal milling.

Surface finish created by milling operations is excellent compared to that from the other operations, and varies between 16 and 500 μin. (0.40–12.5 μm). The tool, which rotates, generally remains at a fixed location, and the workpiece, which is mounted on the bed, is advanced into

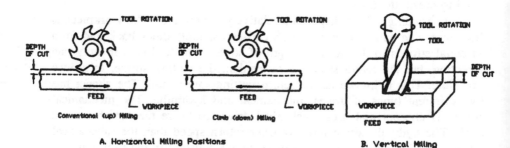

Figure 14.4 Tool rotation, feed, and depth of cut for horizontal and vertical milling positions. (Adapted from Ref. 14.)

the rotating tool. In turning, drilling, and shaping, the tool is advanced into the workpiece and the workpiece remains in a fixed location.

The major problem with the third approach of classifying the processes by operation is that it tends to focus more on specific equipment than on the shape or feature to be generated. It requires a vast knowledge about the processes and their differences; for example, a flat surface can be generated by the various types of milling, shaping or planing, or even by turning (facing). These processes and their tools are quite different, but the basic shapes generated are quite similar.

14.8 MACHINING VARIABLES AND RELATIONSHIPS

There is a wide variety of machining processes, which leads to numerous variables (Appendix 14.1) and relationships (Table 14.2). The key variables related to most of the machining processes are:

1. Cutting speed or velocity (V), ft/min or m/sec
2. Feed (f), in./rev or mm/rev
3. Depth of cut (d), in. or mm

These three variables have a major effect upon the material (or metal) removal rate (MRR), which has a major role in determining the power requirements. In addition, these parameters also have a major effect upon the economics of the processes. Many of the relationships used for the various machining processes are presented in Table 14.2.

The time to produce a part is one of the parameters that has a major impact upon the economics of the process. The time units generally are expressed in terms of minutes, so the velocity units in the ISO-metric system are often m/min rather than m/sec. The time to produce a part can be obtained from:

$$T = \text{Vol/MRR} \tag{14.3}$$

where

       Vol = volume of material removed
      MRR = material removal rate
        T = time

If the depth of cut and length of cut are fixed, as in single-pass cutting, this relationship can be simplified to:

$$T = L/F \tag{14.4}$$

Table 14.2 Machining Relationships

	US-English	ISO-metric
Speed[a]		
Velocity	$V = \pi DN/12$	$V = \pi DN/1000$
RPM	$N = 12V/\pi D$	$N = 1000V/\pi D$
Feed rate[b]		
Turning, drilling	$F = f_r N$	$F = f_r N$
Milling	$F = f_t n_{tr} N$	$F = f_t n_{tr} N$
Shaping	$F = n_s f_s$	$F = f_s n_s$
Cutting time[c]		
Turning, drilling, milling	$T = L/F$	$T = 10L/F$
Shaping	$T = W/F$	$T = W/F$
Metal removal rate[d]		
Turning	$\text{MRR} = \pi Ddf_r N$	$\text{MRR} = \pi Ddf_r N$
	$= 12Vdf_r$	$= 1000Vdf_r$
	$= \pi DdF$	$= \pi DdF$
Drilling	$\text{MRR} = (\pi D^2/4)F$	$\text{MRR} = (\pi D^2/4)F$
	$= 3VDf_r$	$= 250VDf_r$
Milling	$\text{MRR} = WdF$	$\text{MRR} = WdF$
	$= 12Wdf_t n_t V/\pi D$	$= 1000Wdf_t n_t V/\pi D$
	$\text{MRR} = W_c dF$	$\text{MRR} = W_c dF$
	$= 12W_c df_t n_t V/\pi D$	$= 1000W_c df_t n_t V/\pi D$
Shaping	$\text{MRR} = L_c dn_s f_s$	$\text{MRR} = L_c dn_s f_s$
Power calculations		

hp (cutter) = hp_u × MRR

hp (actual) = hp (tare) + hp (cutter)$/E_m$

 hp (tare) = horsepower (hp or kW) to run machine (cutting air)

Shaper calculations and data

$R = 1.6$ for mechanical shaper

$R = 2.0$ for hydraulic shaper

$n_s = 12V/L(1 + 1/R)$ (US-English)

 $= 1000V/L(1 + 1/R)$ (ISO-metric)

Lead calculations

a. Drilling: $l = (D/2) \tan 31 = 0.30D$ (drill angle 118)

b. Milling

   Slab/slot: $l = \sqrt{R^2 - (R - d)^2} = \sqrt{d(D - d)}$

   Face/end $l = D$

---

[a] V = ft/min or m/min, D = in. or mm, N = rpm.

[b] F = in./min or mm/min, f_r = in./rev or mm/rev, n_{tr} = # of cutter teeth, f_t = in./tooth or mm/tooth, n_s = strokes/min, f_s = in /stroke or mm/stroke.

[c] L = in. or mm, W = in. or mm.

[d] MRR = in.³/min or mm³/min.

where

L = length of cut, in. or mm
F = feed rate, in./min or mm/min

The feed rate, F, is a function of the feed (in. or mm/rev) and velocity (ft or m/min), and this expression can be written in two different forms.

a. If the velocity is fixed, the approach is called the *feed-based* approach; the key relationship is:

$$T = L/(N \times f) \tag{14.5}$$

where

$N = V/(\pi \times D)$
V = cutting speed
D = workpiece diameter or tool diameter, depending upon the process

b. If the feed is fixed, the approach is called the *velocity-based* approach; the key relationship is:

$$T = B/V \tag{14.6}$$

where

B = tool cutting path length
V = cutting speed

There are three different approaches to determine the velocity and feed values to use in the cutting time and other metal-cutting relationships:

1. Use the recommended values for V and f found in reference tables.
2. Use graphical relationships that give V and f values as a function of material properties, such as hardness.
3. Use the Taylor tool-life equation parameters to calculate the "optimal" tool life and corresponding velocity. The feed value is found from a reference table.

The first approach would utilize reference data such as that in Table 14.3. The value of N would be calculated from the recommended cutting speed in the reference table, and the value of the feed rate would also be obtained from the reference table. For some operations, such as drilling, general relationships exist for the feed per revolution; that is:

$$f_r = 0.01D \tag{14.7}$$

Table 14.3 Cutting Speed, Feed Rates, and Depth of Cuts for Various Materials for Turning

| Material | Cutting speed for: | | | | Feed rate | |
| | HSS tool materials | | Carbide tool materials | | | |
	Rough	Finish	Rough	Finish	Rough	Finish
Mild steel	130 (40)	200 (60)	300 (90)	600 (180)	0.025–0.080 (0.65–2.00)	0.005–0.030 (0.125–0.75)
Gray cast iron	60 (18)	90 (27)	200 (60)	325 (100)	0.015–0.100 (0.40–2.5)	0.0075–0.040 (0.02–1.0)
Aluminum	300 (90)	500 (150)	800 (240)	1200 (360)	0.004–0.020 (0.10–0.50)	0.0030–0.010 (0.075–0.25)

Units are ft/min (m/min) for velocity and in./rev (mm/rev) for the feed rate. The reference depth of cut for rough cuts was 0.150 in. (4.0 mm) and for finish cuts was 0.025 in. (0.60 mm). Reprinted from Ref. 5, p. 122, by courtesy of Marcel Dekker, Inc.

where

f_r = feed length per revolution
D = diameter of drill

For materials with high machinability, the feed could be doubled; for materials with very poor machinability, the feed would be half of the value calculated by Eq. (14.7).

Note. The expressions and relationships used in the example problems are found in Appendix 14.1 and Table 14.2.

14.8.1 Calculations Involving Turning

Example Problem 14.2. A 1-in.-diameter bar, 6 in. in length, is to be finished turned with a depth of cut of 0.005 in. at a cutting speed of 200 ft/min. The length of pretravel and overtravel is 0.25 in., the unit horsepower is 1.0 hp/in.^{3}/min, the tare horsepower is 0.3 hp, the motor efficiency is 80 percent, and the feed rate used is 0.010 in./rev. Figure 14.5 illustrates the workpiece used.

a. What is the RPM used?
b. What is the horsepower requirements at the cutter?
c. What is the motor horsepower requirements?
d. What is the cutting time?

Solution.

a. $N = V/\pi D$
   $= 200$ ft/min $\times$ 12 in./ft/$(\pi \times 1$ in.$) = 764$ RPM

b. From Table 14.2: hp (cutter) = $hp_u \times$ MRR, and for turning MRR $= \pi D d f_r N$.

   MRR $= \pi \times 1.0$ in. $\times 0.005$ in. $\times 0.010$ in./rev $\times 764$ RPM
   $= 0.12$ in.^{3}/min (finishing cut)
   hp (cutter) $= 0.12$ in.^{3}/min $\times 1.0$ hp/in.^{3}/min
   $= 0.12$ hp

c. hp (actual) = hp (tare) + hp (cutter)/M
   $= 0.30 + 0.12/0.8 = 0.3 + 0.15 = 0.45$ hp

d. time $= L/F = (6.0 + 0.25)/(0.010$ in./rev $\times 754$ RPM)
   $= 6.25$ in./(7.54 in./min) $= 0.83$ min

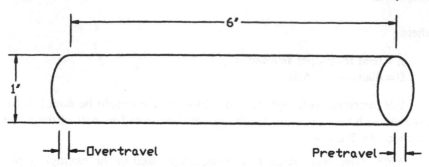

Figure 14.5   Workpiece for Example Problem 14.2.

14.8.2   Calculations Involving Milling

Example Problem 14.3.   An end mill is used to put a 25-mm slot with a depth of 5 mm in a cast iron block with a high-speed cutter. The block is 50 mm wide, 10 mm tall, and 100 mm long. The cutter, a high-speed cutter with a diameter of 25 mm, has four teeth. The pretravel and overtravel combine to a total length of 5 mm. The cut will be made at a feed rate of 0.130 mm/tooth and a cutting speed of 40 m/min. The unit kilowatt power is 0.005 kW/mm³/min. Figure 14.6 indicates the final shape to be produced.

    a.   What is the RPM used?
    b.   What is the length of the lead?
    c.   What is the cutting time?
    d.   What is the metal removal rate?
    e.   What is the power (kW) required at the cutter?

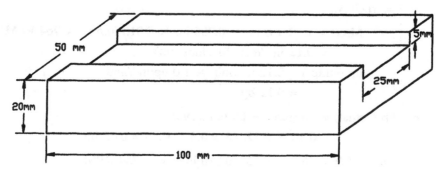

Figure 14.6   Final shape for Example Problem 14.3.

Solution.

a. $N = 1000V/\pi D$

$\quad = 1000 \times 40/(3.14 \times 25)$

$\quad = 510$ RPPM

b. $1 = D$

$\quad = 25.0$ mm

c. $T = L/F$

where

$L = $ piece length + lead + pretravel and overtravel

$\quad = 100$ mm + 25.0 mm + 5.0 mm

$\quad = 130$ mm

$F = f_t n_{tr} N$

$\quad = 0.130$ mm/tooth $\times$ 4 teeth $\times$ 510 rev/min

$\quad = 265$ mm/min

Therefore, the cutting time is:

$T = L/F$

$\quad = 130$ mm/265 mm/min

$\quad = 0.49$ min

d. $\text{MRR} = W_c dF$

$\quad = 25$ mm $\times$ 5 mm $\times$ 265 mm/min

$\quad = 33,125$ mm^3/min

e. kW (cutter) $= hp_u \text{MRR}$

$\quad = 0.005$ kW/mm^3/min $\times$ 33,125 mm^3/min

$\quad = 166$ kW

14.8.3 Calculations Involving Drilling

The calculations involved in drilling are almost identical to those for milling.

Example Problem 14.4. A drill press is used to put a 1-in. hole through a 1-in.-thick aluminum plate. The drill bit has an included standard angle of 118°. The cutting speed is 350 ft/min, and the total pretravel and overtravel is 0.5 in. The unit hp is 0.5 hp/in.^{3}/min. The machinability of aluminum is high. Figure 14.7 illustrates the lead, pretravel, overtravel, and length of cut.

    a.   What is the RPM used?
    b.   What is the length of the lead?
    c.   What is the cutting time?
    d.   What is the metal removal rate?
    e.   What is the horsepower required at the cutter?

Solution.

    a. $N = 12V/\pi D$

$$= 12 \times 350/(3.14 \times 1)$$
$$= 1337.6 \text{ RPM}$$

    b. $l = 0.3D$

$$= 0.3 \times 1$$
$$= 0.3 \text{ in.}$$

    c. $T = L/F$

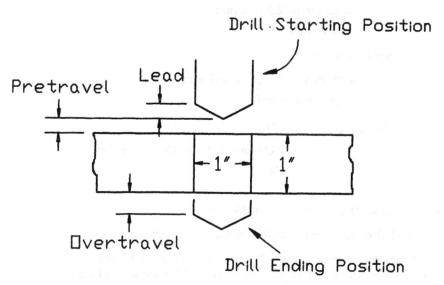

Figure 14.7   Drill positions, lead, pretravel, and overtravel for Example Problem 14.4.

where

L = depth of hole + lead + pretravel and overtravel

= 1 in. + 0.3 in. + 0.5 in.

= 1.8 in.

$F = f_r N$

$f_r = 0.01D$

Note. f_r is doubled because the machinability of aluminum is high.

$f_r = (0.01 \times 1 \text{ in.})2$

= 0.02 in./rev

= 0.02 in./rev × 1337.6 rev/min

= 26.8 in./min

Therefore, the cutting time is:

$T = L/F$

= 1.8 in./26.8 in./min

= 0.07 min

d. MRR = $(\pi D^2/4)F$

= (3.14 × 1 in.^{2}/4) × 26.8 in./min

= 21 in.^{3}/min

e. hp (cutter) = hp_uMRR

= 0.5 hp/in.^{3}/min × 21 in.^{3}/min

= 10.5 hp

14.8.4 Calculations Involving Shaping

Example Problem 14.5. A shaper is used to put a smooth finish on a soft cast iron block with dimensions of 1000 mm × 500 mm. A high-speed steel tool with a recommended speed of 20 m/min and a recommended feed of 3 mm/stroke is used. The pretravel is 10 mm, and the overtravel is 5 mm for the mechanical shaper. The depth of cut is a rough cut of 8 mm.

a. What is the total length of cut?
b. What is the number of strokes per min?
c. What is the feed rate?

  d. What is the metal removal rate?
  e. What is the time per pass?

Solution.

  a. L = piece length + pretravel + overtravel
      = 1000 mm + 10 mm + 5 mm
      = 1015 mm

  b. $$n_s = \frac{1000V}{L(1 + 1/R)}$$
      $$= \frac{1000 \times 20 \text{ m/min}}{1015 \text{ mm } (1 + 1/1.6)}$$
      = 12.1 strokes/min

  c. $F = n_s \times f_r$
      = 12.1 strokes/min × 3 mm/stroke
      = 36.3 mm/min

  d. MRR = $L_t dF$
      = 1000 mm × 8 mm × 36.3 mm/min
      = 290,400 mm^3/min

  e. $T = W/F$
      = 500 mm/36.3 mm/min
      = 13.8 min/pass

14.9 GRAPHICAL-BASED APPROACH TO FEEDS AND SPEEDS

The second approach utilizes the data (8) in Fig. 14.8 in a manner similar to that presented by Schey (4). The advantage of this approach is that it recognizes that the hardness of the material has a significant effect upon the tool life in cutting ferrous materials with high-speed steel (HSS), tungsten carbide (WC), and carbide-coated tools. For finish turning, the cutting speed is 1.25 times the reference cutting speed, and the feed rate is half (0.5 times) the reference feed. The velocity and feed values can be expressed as the product of the adjustment factor and the reference velocity and reference feed values in Fig. 14.8. The feed ranges as well as the cutting speeds are obtained from the chart. The hardness values are Brinell hardness number

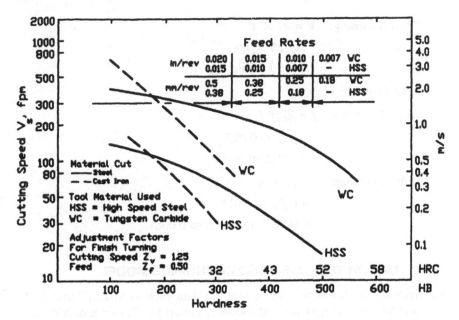

Figure 14.8 Reference speeds and feeds for ferrous materials. (Adapted from Refs. 4 and 13.)

(BHN) values and Rockwell C values. The metric cutting speed is in m/sec and should be converted to m/min for calculating the cutting time values.

Example Problem 14.6.   What is the recommended rough and finishing cutting feeds and speeds for steel with a Brinell hardness of 300? The cutting tool material is high-speed steel (HSS).

Solution.   The adjustment values for Z_v and Z_f are 1.0 for rough turning. For finish turning, the adjustment values for Z_v and Z_f are 1.25 and 0.50, respectively. V_s and f_s can be found from Fig. 14.8 using the material hardness, the metal type, and the type of cutting tool. With a BHN of 300, the material being steel, and an HSS tool being used, Fig. 14.8 can be used to obtain $V_s = 80$ ft/min and $f_s = 0.015$ in./rev.

The cutting speed can be found as follows:

Rough turning: $V = Z_v \times V_s$

$$= 1.0 \times 80$$

$$= 80 \text{ ft/min}$$

Finish turning: $V = Z_v \times V_s$

$\qquad\qquad\qquad =1.25 \times 80$

$\qquad\qquad\qquad = 100$ ft/min

The feed rate is:

Rough turning: $f = Z_f \times f_s$

$\qquad\qquad\quad = 1.0 \times 0.015$

$\qquad\qquad\quad = 0.015$ in./rev

Finish turning: $f = Z_f \times f_s$

$\qquad\qquad\qquad = 0.50 \times 0.015$

$\qquad\qquad\qquad = 0.0075$ in./rev (use 0.008)

14.10 TAYLOR TOOL-LIFE AND ECONOMICS MODEL

The third approach utilizes Taylor tool-life data such as that in Table 14.4. One problem is that it is often difficult to obtain the Taylor tool-life data, for few literature sources present data. The expressions for the total unit cost can be expressed in terms of cutting speed or tool life and are:

$$C_u = Mt_l + MB/V + MBQ[C_t/M + t_{ch}]C^{-1/n}V^{1/n-1} \qquad (14.8)$$

or

$$C_u = Mt_l + (B/C)MT^n + (B/C)MT^n(Q/T)(C_t/M + t_{ch}) \qquad (14.9)$$

Equation (14.8) or (14.9) can be used to determine the unit cost at any tool life (T) or cutting speed (V), and not only the optimal values of T or V. Derivation of these equations and the optimal-tool-life expressions are presented in the appendix to this chapter. The optimal-tool-life expressions obtained for the minimum-cost and maximum-production models are:

i. Minimum-cost model:

$$T = Q(C_t/M + t_{ch}) \times (1 - n)/n \qquad (14.10)$$

ii. Maximum-production model:

$$T = Qt_{ch} \times (1 - n)/n \qquad (14.11)$$

The optimal cutting speeds can be found by using the basic Taylor tool-life equation and solving for V via:

$$V = C/T^n \qquad (14.12)$$

The expressions for the cutting tool path length (B) and the cutting

Table 14.4 Taylor Tool-Life Data

		Taylor tool-life parameters			
Materials		Exponent		Constant	
Tool material	Work material	Range	Typical	US (ft/min)	ISO-metric (m/min)
A. *General tool values*					
HSS	Steel	0.05–0.20	0.10	200	60
Carbide	Steel	0.15–0.35	0.25	500	150
Ceramic	Steel	0.40–0.80	0.50	1600	500
B. *Typical values*[a]					
HSS	Low-C steel		0.10	240	70
HSS	Med-C steel		0.18	170	50
HSS	Low-alloy–med-C		0.11	100	30
HSS	Stainless		0.08	170	50
HSS	Gray cast iron		0.14	75	23
Carbide	Low-C steel		0.28	350	110
Carbide	Med-C steel		0.32	800	240
Carbide	Stainless		0.16	400	120
Carbide	Gray cast iron		0.25	170	50

Values change considerably with the use of cutting fluids and depth of cut. Cutting fluids increase both n and C, whereas depth of cut increases mainly C. Adapted from various sources, including Refs. 6, 8, 9, and 12.

fraction (Q) are found in Table 14.5. These expressions permit the use of the tool-life models for processes such as drilling, milling, and shaping, which are not frequently presented.

14.11 DETAILED METAL-CUTTING PROBLEMS

Example Problem 14.7. A slot is to be milled in a part as illustrated in Fig. 14.9. The following information is about the type of cut and the costs.

Length of cut	10.00 in.	Unit horsepower	1.2 hp/in.^{3}/min
Width of slot	0.50 in.	Machine efficiency	90%
Depth of slot	0.20 in.	Tare horsepower	0.50 hp
Total pretravel and overtravel	0.25 in.		

Table 14.5 Expressions for Cutting Path Length (B) and Fraction of Cutting Time (Q) for Primary Machining (Cutting) Processes

| Cutting process | Expression for cutting path length (B) | | Cutting fraction (Q) | | |
	US-English (ft)	ISO-metric (m)	Typical value	Typical range	Calculation expression
Turning	$\pi DL/12f_r$	$\pi DL/10f_r$	1	0.95–1.0	L_c/L
Drilling	$\pi DL/12f_r$	$\pi DL/10f_r$	1	0.75–0.95	L_c/L
Shaping	$WL/12f,Z$	$WL/10f,Z$	0.67	0.55–0.70	ZL_c/L
Milling	$\pi DL/12n_{tr}f_t$	$\pi DL/10n_{tr}f_t$	0.15	0.03–0.50	$(\Theta/360)(L_c/L)$

Reprinted from Ref. (5) p. 258, by courtesy of Marcel Dekker, Inc.

Labor rate $18.00/hr   Handling time/piece   1.5 min
Overhead rate $30.00/hr   Tool changing time   2.0 min
Average tool cost $25.00

Taylor tool-life exponent = 0.10
Taylor tool-life constant = 200 ft/min

feed rate per tooth = 0.004 in./tooth
number of teeth on cutter = 6
cutter diameter = 4 in.

a. Determine the total length of cut, L.
b. Determine the value of Q.
c. Determine the tool path cutting length B.
d. Determine the tool life for minimum cost.
e. Determine the cutting speed for minimum cost.
f. Determine the total unit cost.

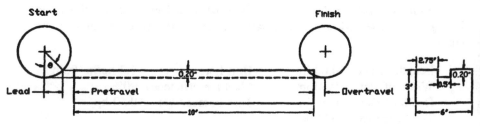

Figure 14.9 Workpiece for Example Problem 14.7.

g. Determine the total unit time.
h. Determine the production rate in pieces/hr.

Solution.

a. $L = L_c$ + (pretravel + overtravel) + lead

lead $= \sqrt{d(D - d)} = \sqrt{0.20 \text{ in.} \times (4.0 \text{ in.} - 0.20 \text{ in.})} = 0.87$ in.

$L = 10$ in. $+ 0.25$ in. $+ 0.87$ in. $= 11.12$ in.

b. From the geometry in Fig. 14.9.

Θ = arcsin(lead/R) = arcsin(0.87/2.0) = 25.8°

Using the expression for Q in Table 14.5:

$Q = (\Theta/360) \times (10/11.12) = 0.064$

c. B = cutting path length, in feet, and using the expression in Table 14.5

$B = \pi DL/12n_t f_t$

$= \pi \times 4$ in. $\times 11.12$ in./(12 in./ft $\times$ 6 teeth $\times$ 0.004 in./tooth)

$= 485$ ft

d. $T = Q(C_t/M + t_{ch}) \times (1 - n)/n$

$M = \$18.00/\text{hr} + \$30.00/\text{hr} = \$48/\text{hr} = \$0.80/\text{min}$

$T = 0.064[\$25.00/(\$0.80/\text{min}) + 2.0 \text{ min}] \times (1 - 0.10)/0.01$

$= 19.15$ min

e. $VT^n = C$

$V = (200 \text{ ft/min})/19.15^{0.1}$

$= 148.9 \text{ ft/min} = 149$ ft/min

f. From Eq. (14.9) or (14.A8) one obtains:

$C_u = Mt_l + MB/V + MBQ(C_t/M + t_{ch})C^{-1/n}V^{1/n-1}$

$= \$.8/\text{min} \times 1.5 \text{ min} + \$.8/\text{min} \times 485 \text{ ft}/149 \text{ ft/min} +$
$\$.8/\text{min} \times 485 \text{ ft} \times 0.064 \times [(\$25/\$.8/\text{min}) + 2 \text{ min}]$
$\times 200^{-1/0.1} \times 149^{1/0.1-1}$

$= \$1.20 + \$2.60 + \$24.832 \times 33.25 \times (200^{-10})(149^9)$

$= \$1.20 + \$2.60 + \$0.29$

$= \$4.09$

g. From Eq. (14.A13):

$$t_p = t_1 + t_c + t_{ch} \times n_t$$

$$= t_1 + B/V + t_{ch} \times QBC^{-1/n}V^{1/n-1}$$

$$= 1.5 \text{ min} + 485 \text{ ft}/(149 \text{ ft/min}) + 2.0 \text{ min} \times 0.064 \times$$
$$485 \text{ ft} \times (200 \text{ ft/min})^{-10}(149 \text{ ft/min})^9$$

$$= 1.5 \text{ min} + 3.26 \text{ min} + 0.022 \text{ min}$$

$$= 4.78 \text{ min}$$

Note. The tool changing time per unit can also be calculated as:

$$t_{ch} \times n_t = t_{ch} \times Qxt_c/T = t_{ch} \times Q \times (B/V)/T$$

$$= 2.0 \times 0.064 \times (485 \text{ ft}/149 \text{ ft/min})/19.15 \text{ min}$$

$$= 0.022 \text{ min}$$

h. The production rate in pieces per hour would be:

$$P = (60 \text{ min/hr})/(4.78 \text{ min/piece}) = 12.6 \text{ pieces/hr}$$

Example Problem 14.8. The material drill combination follows the Taylor tool-life equation form and has the values of:

$$VT^{0.10} = 100 \text{ ft/min}$$

The drill is used to drill 10 holes, each 0.5-in. in diameter, in a plate that is 0.75-in. thick and has a surface of 6 in. × 8 in. The average drill cost is $4.00 per tool life, the labor cost is $9.00 per hour, and the overhead cost is $6.00 per hour. The time to change the drill is 1 min, the time to move from hole to hole for the 10 holes and the time to load and unload the part is a total of 2 min. The total pretravel and overtravel for each hole is 0.10 in., the feed rate is 0.005 in./rev. Use the maximum-production economics model and determine the following:

a. The value for Q for one hole
b. The cutting speed to be used, in ft/min
c. The cutting path length for one hole and for the part
d. The unit cost per piece for drilling
e. The production rate, in pieces per hour

Solution.

a. $Q = L$ (cut)/L (total)

    $= L$ (cut)/$[L$ (cut) + (pretravel + overtravel) + lead]

    where

    L (cut) = 0.75 in. (thickness of part)

    pretravel + overtravel = 0.10 in.

    lead (from Table 14.2) = $0.3D$ = 0.3 × 0.5 = 0.15 in.

    Therefore:

    $Q = (0.75)/[0.75 + 0.10 + 0.15] = 0.75/1.00 = 0.75$

b. First find the tool life, and then find the cutting speed. For the maximum-production model:

    $T = Q \times t_{ch} \times (1 - n)/n = 0.75 \times 1.0 \times (1 - 0.10)/0.10$

    $= 6.75$ min

    Therefore:

    $V = C/T^n = 100/(6.75)^{0.10} = 82.6$ ft/min

c. $B = \pi DL$ (total)/$12f_r$

    $= \pi \times (0.50 \text{ in.}) \times (1.0 \text{ in.})/(12 \text{ in./ft} \times 0.005 \text{ in./rev})$

    $= 26.2$ ft

    If 10 holes are drilled, the B (total) = 10 × B = 262 ft.

d. If one uses Eq. (14.8):

    $$C_u = Mt_l + MB/V + MBQ(C_t/M + t_{ch})C^{-1/n}V^{1/n-1}$$

    where

    M = ($6.00/hr + $9.00/hr)/60 min/hr = $0.25/min

    t_l = 2.0 min

    B = 262 ft

    V = 82.6 ft/min

    Q = 0.75

    C_t = $4.00

    n = 0.10

    t_{ch} = 1.0 min

    C = 100 ft/min

Thus:

$$C_u = \$0.25/min \times 2\ min + \$0.25/min \times 262\ ft/82.6\ ft/min$$
$$+ \$0.25/min \times 262\ ft \times 0.75 \times (\$4.00/\$0.25/min +$$
$$1.0\ min) \times 100\ ft/min^{-1/0.10} \times 82.6\ ft/min^{1/0\ 1-1}$$

$$= \$0.500 + \$0.793 + (49.125 \times 17 \times 0.00179)$$

$$= \$0.500 + \$0.793 + \$1.495 = 2.788 = \$2.79/piece$$

e. To find the production rate, one must first determine the unit time per piece:

$$t = t_l + t_c + t_{ch}n_t$$

$$= 2.0\ min + 262\ ft/82.6\ ft/min + 1.0\ min \times Qxt_c/T$$

$$= 2.0\ min + 3.17\ min + 1.0 \times (0.75 \times 3.17\ min/6.75\ min)$$

$$= 2.0\ min + 3.17\ min + 0.35\ min$$

$$= 5.52\ min/piece$$

$$P = (60\ min/hr)/(5.52\ min/piece)$$

$$= 10.9\ pieces/hr$$

14.12 NONMECHANICAL (NONTRADITIONAL) MACHINING

The information presented on machining has only been on mechanical machining. The other methods of metal removal are classified as nonmechanical or nontraditional machining. There are four basic nonmechanical processes (7):

1. *Chemical milling*: The metal is removed by the etching reaction of chemical solutions on the metal.
2. *Electrochemical machining*: Uses the principle of metal plating in reverse: the workpiece, instead of being built up by the plating process, is dissolved in a controlled manner by the an electric current.
3. *Electric discharge machining and grinding*: Cuts or erodes the metal by means of high-energy electrical discharges.
4. *Lasers*: A concentrated and amplified source of monochromatic light used for cutting.

Nonmechanical machining is not as widely used as mechanical machining. It is used mainly for special manufacturing problems, such as machining extremely hard materials, that would be difficult to accomplish with traditional mechanical machining.

14.13 SUMMARY

The machining processes are the most widely used of all the manufacturing processes, since they are both a primary shaping process as well as a secondary finishing shaping process. The large amount of waste in the machining process is the major disadvantage of the process. Machining processes are used to improve the surfaces of castings and bulk deformation products because they have excellent surface finish and tolerances capabilities. The primary variables for machining processes are cutting speed, feed, and depth of cut. These parameters have a major impact upon the tool life, the process economics, and the power requirements.

The classification of machining processes have generally been done by describing each process and then describing the shapes or features that can be generated by that process. It would be more desirable to classify the shapes or features into general categories and then to determine the processes that can generate those features. This type of system would help the design engineer who knows the shape he wants to generate to understand the processes that can generate the desired shape.

14.14 EVALUATIVE QUESTIONS

1. What are the primary reasons for selecting machining over casting or bulk deformation?

2. What are the three different types of chips generated by machining processes?

3. What are the two main types of tool wear considered that follow the Taylor tool-life equation?

4. What are the three main variables in most machining processes?

5. What are the four types of surfaces generated by machining processes?

6. What are the two types of metal-cutting models used for the analysis of forces?

7. A material–tool combination follows the Taylor tool-life relationship of:

$$VT^{0.3} = 1000 \text{ ft/min}$$

  a. If the cutting speed is increased by 20 percent, what is the percentage decrease in tool life?
  b. If the exponent is decreased from 0.3 to 0.1 and the cutting speed is increased by 20 percent, what is the percentage decrease in tool life?
     (a. 46 percent decrease; b. 84 percent decrease)

8. A 2-in.-diameter bar, 6 in. in length, is turned on a lathe with a depth of cut of 0.100 in.
    a. If the recommended cutting speed is 100 ft/min, what RPM is to be used? (191)
    b. If the RPM used is 200 and the feed is 0.020 in./rev., what is the metal removal rate? (2.51 in.³/min)
    c. If the RPM is 200 and the unit horsepower is 1.8 hp/in.³/min, what are the horsepower requirements at the cutter and for the motor if the tare is 0.3 hp and the motor efficiency is 80 percent? (4.52 hp, 5.95 hp)
    d. What would be the cutting time? (1.5 min)

9. A 1-in. hole is to be drilled through a steel plate 1 in. × 6 in. × 10 in. The recommended cutting speed is 80 ft/min, the feed is 0.020 in./rev, and the unit horsepower is 1.5 hp/in.³/min.
    a. What is the cutting time? (12.7 sec)
    b. What is the horsepower required at the cutter? (7.2)

10. A 1-in. hole is to be drilled through a cast iron block 6 in. × 4 in. × 1 in. The recommended cutting speed is 150 ft/min, the unit horsepower is 1.2 hp/in.³/min, the feed is 0.02 in./rev, and the drill has the standard angle of 118°.
    a. What is the RPM? (573)
    b. What is the metal removal rate? (9 in.³/min)
    c. What is the horsepower required at the cutter? (10.8 hp)
    d. What is the normal time to drill the hole? (0.113 min)

11. The cast iron block of Evaluative Question 3 is to have the surface finished by removing 0.100 in. from one 4 × 6 surface. The piece is put into the shaper so the tool will travel over the 6-in. length and feed across the 4-in. side. The feed per stroke is 0.015 in., the unit horsepower is 1.4 hp/in.³/min, and the recommended speed is 100 ft/min. Assume a mechanical shaper and pretravel and overtravel of 1 in.
    a. What is the number of strokes per minute? (105)
    b. What is the metal removal rate? (0.945 in.³/min)
    c. What is the horsepower required at the cutter? (1.32 hp)
    d. What is the time to cut the piece? (2.54 min)

12. If a steel has a Brinell hardness of 350 (350 BHN) and throwaway carbide tool inserts are used, then:
    a. What are the recommended cutting speed and feed for a roughing cut?
    b. What are the recommended cutting speed and feed for a finishing cut?

13. A 6-in.-diameter SAE 1020 steel bar 30 in. long is to be machined on a lathe under the following conditions:

Tool cost data: $4/insert, 4 edges per insert

Feed: 0.020 in./rev	Tool changing time: 0.50 min
Depth of cut: 0.100 in.	Tool handling time: 1.00 min
Labor rate: $12/hr	Tool-life exponent: 0.25
Machine overhead rate: $18/hr	Tool-life constant: 3000 ft/min
$Q = 1.0$	

   a. For the minimum-cost case, determine the cutting speed, the tool life, the unit time per piece, and the cost per piece.
   b. For the maximum-production case, determine the cutting speed, the tool life, the unit time per piece, and the cost per piece.
   (a) 1813 ft/min; 7.5 min; 2.39 min; and $1.37/piece, $B = 2356$ ft
   (b) 2711 ft/min; 1.5 min; 2.16 min; and $1.65/piece

14. The material drill combination follows the Taylor tool-life equation of:

$$VT^{0.10} = 100 \text{ ft/min}$$

The drill is used to drill 10 holes, each $\frac{1}{2}$ in. in diameter, in a plate that is 6 × 8 in. on the surface and $\frac{3}{4}$ in. thick. The drill cost is $4.00 per use (average tool cost), the labor cost is $9.00/hr, and the overhead cost is $6.00/hr. The time to change the drill is 1 min. The time to move from hole to hole for the 10 holes and the time to load and unload the part is a total of 2 min. The total pretravel and overtravel for each hole is 0.10 in., and the feed rate is 0.0050 in./rev. Use the maximum-production economics model and determine the following:
   a. The value of Q (for one hole)
   b. The cutting speed, in ft/min
   c. The cutting path length for one hole (feet) and for 10 holes
     (a. 0.75, b. 82.6, c. 26.2 ft and 262 ft)

REFERENCES

1. Edwards, L., and Endean, M. Manufacturing with Materials, Butterworths, London, 1990.
2. Roberts, A. D., and Lapidge, S. C. Manufacturing Processes, McGraw-Hill, New York, 1977, p. 58.
3. Boothroyd, G., and Knight, W. A. Fundamentals of Machining and Machine Tools, 2nd ed., Marcel Dekker, New York, 1989.

4. Schey, J. A. Introduction to Manufacturing Processes, 2nd ed., McGraw-Hill, New York, 1987, pp. 505–506.
5. Creese, R. C., Adithan, M., and Pabla, B. S. Estimating and Costing for the Metal Manufacturing Industries, Marcel Dekker, New York, 1992, pp. 115–147, 253–265.
6. Niebel, B. W., and Draper, A. B. Product Design and Process Engineering, McGraw-Hill, New York, 1974.
7. U.S. Department of Commerce. International Trade Administration Office of Industrial Resource Analysis Division. Investment Casting: A National Security Assessment. December 1987, p. 111.
8. Shaw, M. C. Metal Cutting Principles, Oxford University Press, Oxford, England, 1984, pp. 235–240, 556–585.
9. Drozda, T. J., and Wicks, C., eds., Tool and Manufacturing Engineers Handbook, Volume 1: Machining, Society of Manufacturing Engineers, Dearborn, MI, 1983, pp. 1-45–1-47.
10. Metals Handbook, Volume 16: Machining, 9th ed., ASM International, Metals Park, OH, 1989, p. 944.
11. Bolz, R. W. Production Processes, Conquest Publications, Novelty, OH, 1974.
12. Amstead, B. H., Ostwald, P. F., and Begeman, M. L. Manufacturing Processes, 8th ed., Wiley, New York, 1987, pp. 467–471.
13. Metals Handbook—Desk Edition, ASM International, Metals Park, OH, 1985, pp. 27-30–27-54.
14. Todd, R. H., Allen, D. K., and Alting, L. Manufacturing Processes Reference Guide, Industrial Press, New York, 1994, pp. 7, 49.

INTERNET SOURCES

Prices and parameters of metal-cutting machines: *http://www.machinetools.com/metcut.html*

Machining data for cast iron, austempered ductile iron: *http://iams.org/isd/news letters/machining/vol1issue2.htm*

Milling process: *http://www.iprod.auc.dk/procesdb/milling/start.htm*

Turning process: *http://www.iprod.auc.dk/procesdb/turning/start.htm*

Drilling process: *http://www.iprod.auc.dk/procesdb/drilling/start.htm*

Introduction to abrasive jet machining, ultrasonic machining, and electrochemical machining: *http://www.cemr.wvu.edu/~imse304/*

APPENDIX 14.A DERIVATION OF OPTIMAL TOOL-LIFE VALUES FOR METAL-CUTTING ECONOMICS MODELS

Appendix 14.A.1 Metal-Cutting Economics Model

Model base: Basic Taylor tool-life equation:

$$VT^n = C$$

Appendix 14.A.2 Nomenclature and Terms

Symbol	Description and Units
t_l	Handling (nonproductive) time (min) per unit
t_{ch}	Tool changing time (min) per unit
t_c	Machining (cutting) time (min) per unit
t_p	Total unit production time (min)
M	Total machine and operator rate (including overheads), in \$/min
C_t	Average tool cost (average of original tool cost + resharpening costs), in \$/tool
V	Cutting speed (velocity), in ft/min or m/min
T	Tool life, in min
n	Taylor tool-life exponent
C	Taylor tool-life constant, in ft/min or m/min
B	Cutting path length, in ft or m
n_t	Tools per unit
C_u	Total unit cost, \$/unit
Q	Fraction of cutting time cutting edge is actually cutting
L	Length of cut, including pretravel, overtravel, and lead; in or mm
L_c	Length of cut when tool is actually cutting, in in. or mm
d	Depth of cut, in in. or mm
W	Width of piece for shaping, in in. or mm
W_c	Width of cutter
D	Diameter of work or cutter, in in. or mm
f_r	Feed per revolution, in in./rev or mm/rev
f_t	Feed per tooth for milling, in in./tooth or mm/tooth
f_s	Feed per stroke for shaping, in in./stroke or mm/stroke
Θ	Angle of contact of milling cutter, in degrees
R	Return speed/cutting speed ratio for shaper
Z	$R/(R + 1)$
n_{tr}	Number of teeth per revolution of milling cutter
n_s	Number of strokes per minute for shaper
N	Revolutions/minute

Appendix 14.A.2 continued

Symbol	Description and Units
F	Feed rate, in in./min or mm/min
MRR	Metal removal rate, in in.^{3}/min or mm^3/min
hp	Unit power requirements, in hp or kW
hp_u	Unit power requirements, in hp/in.^{3}/min or kW/mm^3/min
E_m	Machine efficiency (decimal)
l	Lead: extra distance cutter must move (in. or mm)

Appendix 14.A.3   Minimum-Cost Economics Model

Handling cost $\qquad M \times t_l$ $\qquad\qquad\qquad\qquad$ (14.A1)

Cutting cost $\qquad\quad M \times t_c$ $\qquad\qquad\qquad\qquad$ (14.A2)

Tool cost $\qquad\qquad C_t \times n_t$ $\qquad\qquad\qquad\qquad$ (14.A3)

Tool changing cost $\quad M \times t_{ch} \times n_t$ $\qquad\qquad\quad$ (14.A4)

The total unit cost is the sum of these four terms:

$$C_u = Mt_l + Mt_c + n_t(C_t + Mt_{ch}) \qquad\qquad (14.A5)$$

However:

$$t_c = B/V \qquad\qquad\qquad\qquad\qquad\qquad\qquad (14.A6)$$

$$n_t = Qt_c/T \qquad\qquad\qquad\qquad\qquad\qquad\quad (14.A7)$$

where

Q = fraction of cutting time that tool is actually cutting

Thus:

$$C_u = Mt_l + MB/V + MBQ(C_t/M + t_{ch})C^{-1/n}V^{1/n-1} \qquad (14.A8)$$

Setting the derivative to zero and solving for V one obtains:

$$V = \left[\frac{1}{[Q(C_t/M + t_{ch})]} \times (n/[1-n]) \right]^n \times C \qquad (14.A9)$$

Using $VT^n = C$ and solving for T one obtains:

$$T = Q(C_t/M + t_{ch}) \times (1-n)/n \qquad\qquad (14.A10)$$

The expression for the unit cost can be expressed in terms of the tool life (because the tool life is easier to calculate) as:

$$C_u = Mt_l + (B/C)MT^n + (B/C)MT^n(Q/T)(C_t/M + t_{ch}) \qquad (14.A11)$$

14.A.4 Maximum-Production Model

$$t_p = t_l + t_c + t_{ch} \times n_t \tag{14.A12}$$

The values of t_c and n_t can be expressed in terms of V, and thus:

$$t_p = t_l + B/V + QBt_{ch} \times C^{-1/n} \times V^{1/n-1} \tag{14.A13}$$

Setting the derivative to zero and solving for V and T one obtains:

$$V = \left[\frac{1}{Qt_{ch} \times (n/(1-n))} \right]^n C \tag{14.A14}$$

$$T = Qt_{ch} \times (1-n)/n \tag{14.A15}$$

The value of T from Eq. (14.A15) can be used in Eq. (14.A11) to find the unit cost directly without finding V. The unit cost can be found with the value of V and using Eq. (14.A8). The previous models for metal-cutting economics did not include the Q value, which prevented the extension to drilling, milling, and shaping operations.

15
Joining Processes, Design, and Economics

15.1 INTRODUCTION TO JOINING

The processes of casting, deformation processing, and machining are concerned primarily with the production of components, whereas joining is concerned primarily with the assembly of the components into subassemblies or final products. The joining processes can be divided into three major categories: fasteners, welding, and adhesives. These categories are extremely broad and contain a vast amount of information; but for an initial manufacturing processing course, they are grouped together because all of them focus upon the assembly of components or parts into subassemblies or final products.

The instances when joining is the preferred method of producing the desired shape occur when:

1. The product cannot be made as a single piece.
2. The product is more economical as an assembly (uses several components having different required product properties)
3. Products must be disassembled.
4. Products are easier to transport disassembled.

The main design advantage of assemblies is that the different components can have vastly different properties. For example, a cooking pot needs to have good conduction so the water can boil, but one does not want the handle to get hot. Another example would be a milling cutter, where the teeth need to have good wear resistance but the cutter body must also have good ductility and impact resistance. Products such as bookcases and metal shelving, which have large surface areas, would not be practical as a casting, forging, or machined part.

Here is a more detailed definition of the three major joining process categories and their differences:

Table 15.1 Major Joining Process Groups and Subgroup Examples

Fasteners	Adhesive processes	Cohesive processes
Nonpermanent	Metallic adhesives	Liquid
Threaded	Brazing	Arc welding
Pins	Soldering	Resistance welding
Special purpose	Braze welding	Electron beam
Permanent	Organic bonds (gluing)	Solid
Rivets	Glue	Friction welding
Seam	Epoxy	Diffusion bonding
Shrink fits		Forge welding

Fasteners: Mechanical devices that join materials by means of clamping forces, pressure, or friction, and do not involve molecular bonding between the surfaces as the primary bonding force. Some
common fasteners are nails, screws, nuts and bolts, pins, and paper
clips.

Cohesion (welding): The joining of two or more pieces of material by
means of heat, pressure, or both, with or without a filler metal, to
produce bonding through fusion or recrystallization. The force of
attraction in the bond is primarily cohesion. A few of the cohesion
processes are arc welding, laser welding, diffusion bonding, and
forge welding.

Adhesion (gluing): The joining of two or more pieces of material by
the forces of attraction between the adhesive and the materials being
joined (adherends). Gluing processes depend primarily upon adhesive bonding. It includes processes such as brazing, soldering, and
epoxy bonding.

Fasteners do not involve molecular bonding, whereas cohesive and
adhesive bonding both involve molecular bonding. The welding process is
primarily cohesive, whereas the gluing processes involve primarily adhesive
bonding. A general overview of the joining processes is presented in Table
15.1.

15.2 FASTENER CATEGORIES AND DESCRIPTIONS

Fasteners include an extremely large group of items. They are divided into
two major categories with respect to assembly; see Table 15.2. A brief description (1) of some of the more common fasteners follows the major
categories.

Table 15.2 Major Fastener Categories and Some Fastener Types

Disassembly permitted
A. Threaded fasteners
   1. Nuts and bolts
   2. Screws
B. Pins
   1. Machine pins (straight, dowel, etc.)
   2. Radial locking pins
C. Washers and retaining rings
   1. Washers: distribute compressive stresses over a wider area than the bolt face
   2. Retaining rings: provide a removable shoulder to locate, retain, or lock
      components
D. Special-purpose fasteners, clips, etc.
   1. Quick opening: tends to operate against spring pressure to snap into place
   2. Self-sealing: for leak joints
   3. Spring clips
Permanent assembly
A. Rivets (usually stronger and more economical than threaded fasteners)
B. Metal stitching
C. Staples
D. Seam joints

15.2.1 Common Fasteners

There is a wide variety of fasteners, and this is a major industry itself. Nearly every usable manufactured product contains fasteners, and the different combinations of fasteners approach millions in number. Some of the more commonly used fasteners are described in an abbreviated next; more details can be found in the Standard Handbook of Fastening and Joining (1).

15.2.1.1 Threaded Fasteners

A threaded fastener system consists of two parts, an external (male) thread, such as a screw or a bolt, and a matching internal (female) thread, like a nut, an insert, or a tapped hole. The most common threaded fastener system is the nut and bold. A screw is similar to a bolt, except the external threads have a self-drilling design and do not require a nut to fasten or join, since the part acts as the matching internal thread.

15.2.1.2 Pins

There are various types of pins. Straight pins and dowel pins are the basic design for a pin. Straight pins are usually machined from bar stock and

chamfered at both ends or left square. They must have a smooth finish and
be free of defects.

Dowel pins are the same as straight pins, except straight pins are designed to barely fit into an opening and to require some force to be inserted.
Dowel pins can be applied with little or no force. They are used for machines, jigs, and fixtures.

15.2.1.3   Washer and Retaining Rings

Retaining rings are fastening devices that fit in a slotted groove for positioning or limiting the movement of a part or an assembly. Retaining rings
have to be made of a material that can be deformed elastically and spring
back to original shape.

Washers are used with a threaded fastener. A washer is used to distribute compressive stresses over a wider area than the bolt face. Some
washers can provide a good surface for good torque control. Overall, washers can be used to improve joint efficiency and design.

15.2.1.4   Riveting

Rivets are more economical than threaded fasteners, because they are produced in tremendous quantities and can be applied at high speed. A rivet
can be applied to any hole as long as the shaft of the rivet is long enough.
Rivets have tensile strength and fatigue strength lower than those of a bolt,
but they do have higher compression and shear strength than most other
fasteners.

15.2.2   Bolt Fastener Design Considerations

There is a fastener industry, and billions of fasteners are produced yearly.
However, since the cost of an individual fastener is so small, the industry
does not receive the recognition that it deserves. Although the industry is
very broad, each fastener type generally has considerable design effort. For
example, in the design of an automobile engine, the top part (head) must be
fastened to the main part (block). Threaded bolts are used to fasten the head
to the block. The design of the bolt is important and considers not only
factors such as the strength of the bolt, but also the desired failure mode of
the bolt. A failure can occur in one of three ways:

1. Tensile failure of the bolt
2. Shear failure of the bolt (external threads)
3. Shear failure of the hole (internal threads)

The desired mode of failure is a tensile failure of the bolt, for a shear failure of the bolt results in a scrapping of the engine. (Although the hole could be enlarged and a new fastener used, this would result in a nonstandard part and impose significant potential liability and service problems). To control the failure mode, the bolt length is controlled so that the bolt will fail in tension rather than shear. This is controlled by making the length of thread engagement greater than what would allow shear failure of the threads. This is done by designing the tensile load to be less than the shear load for failure. That is,

tensile load $\leq$ shear load

$$A_t S_t \leq A_s S_s \times N \times L \tag{15.1}$$

$$L \geq 1/N \times A_t/A_s \times S_t/S_s \tag{15.2}$$

$$L \geq P \times A_t/A_s \times S_t/S_s \tag{15.3}$$

where

L = length of engagement
P = pitch of threads
N = number of threads = $1/P$
A_t = tensile area of bolt
A_s = shear area of bolt or hole
S_t = UTS of bolt
S_s = USS of bolt or hole

The length of engagement for both the bolt and the hole must be calculated, and the greatest length must be used to ensure that tensile failure occurs in the fastener. The tensile and shear strength values of typical threaded fasteners are presented in Table 15.3, along with the threaded area values for two different standard bolt sizes.

Example Problem 15.1. What should be the minimum length of engagement of a $\frac{1}{2}$-13, grade 5 bolt to hold a cast iron block and cast iron head together? The $\frac{1}{2}$ refers to the nominal diameter of the bolt, which is $\frac{1}{2}$ in., and the 13 refers to the number of threads per inch. The pitch is the reciprocal of the number of threads per inch, or $\frac{1}{13}$.

Solution. The length of engagement can be determined from using the data in Table 15.3 and Eq. (15.3): For a $\frac{1}{2}$-13, grade 5 bolt with a cast iron block, the length values would be:

Table 15.3 Data for Example Problem 15.1

Ultimate tensile strength and ultimate shear strength values					
Bolt material	psi[a]	MPa[a]	Nonbolt material	psi	(MPa)
Gr 5 UTS	120,000	(826)	Cast iron UTS	27,000	(186)
Gr 5 USS	92,000	(634)	Cast iron USS	33,500	(231)
Gr 1 UTS	60,000	(413)	PM aluminum UTS	23,500	(159)
Gr 1 USS	36,000	(248)	PM aluminum USS	13,900	(96)
			SAE AA 6061-T4 UTS	30,000	(207)
			SAE AA 6061-T4 USS	24,000	(165)
			SAE 903 zinc UTS	41,000	(283)
			SAE 903 zinc USS	31,000	(214)
			SAE 1010 HR UTS	47,000	(324)
			SAE 1010 HR USS	35,400	(244)
			SAE 1010 CR UTS	53,500	(369)
			SAE 1010 CR USS	39,800	(274)

Thread area values for different bolt sizes			
Bolt size UNC Class 2	A (tension) (in.^{2})	A (shear-bolt) (in.^{2}/thread)	A (shear-hole) (in.^{2}/thread)
1-8	0.6060	0.2070	0.2918
$\frac{3}{4}$-10	0.334	0.1214	0.1723
$\frac{1}{2}$-13	0.1419	0.0599	0.0866
$\frac{3}{8}$-16	0.0775	0.0348	0.0518
$\frac{1}{4}$-20	0.0318	0.0184	0.02699
5-40 (0.125)	0.00796	0.0040	0.00619
ISO 6g/6H (medium fit)	(mm^2)	(mm^2/thread)	(mm^2/thread)
24-3	353	121.4	165.0
20-2.5	245	83.2	113.5
14-2	115	44.85	62.07
10-1.5	58	23.37	32.22
6-1	20.1	8.646	12.19
5-0.8	14.2	5.659	7.999
4-0.7	8.78	3.831	5.438
3-0.5	5.03	1.952	2.774

Adapted from Ref. 2.
[a] 1000 psi = 6.895 MPa.

   a. Shear of bolt (external threads):

$$L = \tfrac{1}{13} \times 0.1419/0.0599 \times 120{,}000/92{,}000 = 0.238 \text{ in.}$$

   b. Shear of hole (internal threads):

$$L = \tfrac{1}{13} \times 0.1419/0.0866 \times 120{,}000/33{,}500 = 0.451 \text{ in.}$$

Therefore, the length of engagement of the bolt should be greater than or equal to the greatest length calculated, which is 0.451 in. The cast iron will shear before the bolt will shear, but the design is to make the bolt fail in tension before the cast iron will shear. The high tensile strength of the bolt requires a large length of engagement to avoid the shearing of the threads in the block. The clamping force load that will be supported by the bolt is:

$$P = A_t \times UTS_t$$
$$= 0.1419 \text{ in.}^2 \times 120{,}000 \text{ psi}$$
$$= 17{,}028 \text{ lb} \tag{15.4}$$

   Example Problem 15.2. What should be the minimum length of engagement of a 14-2, grade 5 bolt to hold a cast iron block and cast iron head together? The 14 refers to the nominal diameter of the bolt, which is 14 mm, and the 2 refers to the pitch, which is 2 mm. The number of threads per millimeter is $\tfrac{1}{2}$.

   Solution. The length of engagement can be determined from using the data in Table 15.3 and Eq. (15.3): For a 14-2, grade 5 bolt with a cast iron block, the length values would be:

   a. Shear of bolt (external threads):

$$L = \tfrac{1}{2} \times 115/44.85 \times 826/634 = 1.67 \text{ mm}$$

   b. Shear of hole (internal threads):

$$L = \tfrac{1}{2} \times 115/62.07 \times 826/231 = 3.31 \text{ mm}$$

Therefore, the length of engagement should be at least 3.31 mm.

   Torque is the foot-pounds (newton-meters) of moment that should be applied when tightening the bolt. To achieve the maximum clamping force that the threaded fastener can withstand, the bolt must be torqued correctly. If the bolt is torqued too much, then there is a chance of stripping the threads or breaking the bolt. If it is torqued too little, there is a possibility of fatigue and/or vibration failure. To achieve a certain clamping force, the torque can be found using the equation derived from the torque tension relationship. The torque can be found by:

$$T = K \times D \times P \times K2 \tag{15.5}$$

where

> T = installation torque, lb-ft (N-m)
> K = torque coefficient (friction)
> D = nominal bolt diameter, in. (mm)
> P = clamp load objective, lb (N)
> $K2 = \frac{1}{12}$ to convert inches to feet or 1/1000 to convert mm to meters

The calculations performed earlier in this section found all the variables for torque except the torque coefficient. The torque coefficient will vary for bolts with different finishes used for protection and lubrication. A zinc-and-cadmium finish is a popular plating finish used to reduce the torque coefficient. The torque coefficient of 0.20 is an estimate for unlubricated and dry fasteners; it is approximately 0.15 for plated finishes. The use of additional oils and lubricants can vary the K factor from 0.07 to 0.20 (4).

What most people do not realize is that when a bolt is torqued, the reading from a torque wrench also includes the force to overcome friction. The torque coefficient changes after the continued use of a bolt. As few as five installations can alter the torque coefficient characteristics. The torque coefficient is also increased due to rusting and dirt particles getting on the bolt. Since approximately 15 percent of the total tightening torque produces the clamp load, a 5 percent variation in friction could produce a 30 percent variation in clamp load.

Example Problem 15.3.   What should the torque of a plated bolt with a clamping force load of 7500 lb and a diameter of $\frac{9}{16}$ in. be if the torque coefficient is 0.15?

Solution.   The torque can be found using Eq. (15.5):

$$T = K \times D \times K2 \times P$$

where

> T = installation torque, lb-ft
> K = torque coefficient (friction)
> D = nominal bolt diameter, in.
> P = clamp load objective, lb
> $K2 = \frac{1}{12}$
> $T = 0.15 \times \frac{9}{16} \times \frac{1}{12} \times 7500$
> $\quad = 52.7$ lb-ft

The bolt should be torqued at 52.7 lb-ft. The torque coefficient can be greater, depending on the condition of the bolt.

15.3 ADHESIVE BONDING

There are two major types of adhesives, the *organic* adhesives, such as epoxies and glues, and the *metallic* adhesives, such as brazing, soldering, and braze welding. The strength test for bonding for adhesives is generally cleavage or peel (exaggerated cleavage), since the strength of most adhesives is not high. The type of design loading for failure prevention is generally shear.

The general advantages of adhesives over fusion (welding) bonding are:

1. Dissimilar metal/materials can be joined relatively easy.
2. Joining is at low temperatures.
3. Easy to bond thin-gage materials.
4. Used as sealant.
5. Used as insulator.
6. Easy to apply.
7. Cost is generally low.

The disadvantages of the use of adhesives are:

1. Low joint strength (3000–6000 psi or 20–40 MPa for most organic adhesives, and 20,000–60,000 psi or 140–400 MPa for most metallic adhesives).
2. Low peel strength.
3. Generally restricted to room temperature applications.
4. Generally, organic adhesives are poor conductors.
5. Curing time and temperature may be required, which increases production time and costs.

15.3.1 Epoxy and Gluing Adhesives

There is a wide variety of adhesives, and the types of bonding can be different. The different types of bonding that may occur in an adhesive bond between the adherend and the adhesive are:

1. Secondary bonding (electrostatic or hydrogen)
2. Mechanical bonding (interlocking, especially in porous materials)
3. Diffusion (across interface)
4. Chemical reaction (primary bonding—ionic or covalent)

Since there are several different types of bonding, a successful single theory has not been developed. Thus, the type of bond that forms and the strength of the bond depends upon the properties of both the adhesive and the adherends.

The application of the adhesive follows a general procedure, to ensure that good bonding occurs. The procedure consists of the following steps.

1. *Prepare surface.*
   a. The surface must be clean from dirt, grease, and oxides.
   b. The surface must be reactive and not oxidized.
   c. A large surface area is needed, because the adhesive strengths are relatively low.
2. *Prepare adhesive*: Adhesives often have a short shelf life and must be prepared immediately prior to use.
3. *Apply adhesive.*
   a. Adhesives are generally applied with a brush or spray.
   b. Factors in the application of the adhesive are:
      i. Number of coats
      ii. Time between coats
      iii. Rate of spread of adhesive
4. *Apply pressure*: Pressure is generally applied to help adhesive make good contact with the surfaces of the adherends.
5. *Cure*: Primary factors in the curing of the adhesive are the time and temperature. High temperatures and long curing times hinder application in high-production applications.

15.3.2 Brazing and Soldering

Brazing and soldering are two methods of adhesive bonding that are often incorrectly classified as welding processes, since they involve metallic bonding. According to the American Welding Society (3), the definitions of these two processes are:

Brazing: The filler metal has a liquidus above 450°C (840°F) and below the solidus of the base metal; and the filler metal is distributed by capillary attraction.

Soldering: The filler metal has a liquidus below 450°C (840°F) and below the solidus of the base metal; and the filler metal is distributed by capillary attraction.

The filler metals for brazing contain metals such as copper, silver, and gold, and the filler metals for soldering contain metals such as lead, tin, and cadmium. The general procedure for brazing and soldering is as follows.

1. *Good fit and proper clearances*: The joint clearance is typically 0.001–0.010 in., or 0.025–0.25 mm, depending upon the base metal, flux, and filler metal.

2. *Clean surfaces*: The surfaces must be clean of dirt, grease, or oxides.
3. *Flux*: The purposes of the flux are:
   a. Decompose oxides
   b. Protect the surface during the heating
   c. Reduce the surface tension between the base metal and filler metal so flow is better
4. *Assemble and support*: The pieces must be supported during the heating stage, and the filler is either preplaced or applied during the heating cycle.
5. *Heat and flow of alloy*: Must have controlled heating to melt filler metal and have capillary flow into joints.
6. *Cooling and cleaning of assembly*: Must avoid residual stresses from heating and cooling, and must remove fluxes or they may cause corrosion.

15.3.3 Braze Welding

Braze welding is a process that is neither brazing nor welding. It is not brazing because it does not depend upon capillary flow, and it is not welding because it does not melt the base metal and have fusion. It is an adhesive process, but it looks like a weld in appearance and uses a brazing rod for filler metal. This is why it is called braze welding. The primary application has been to repair relatively brittle metals, such as cast iron, which cannot be welded under most circumstances. Since it looks more like a weld in appearance, the design of braze welds follows those of welds rather than those of most adhesives.

15.3.4 Adhesive Design Considerations

The basic joint for a braze weld is a lap joint or a butt joint. Butt joints are not used if the thickness of the lap joint is too thick for its application. Soldering also uses the lap joint but tries to avoid using the butt joint. The strength of a lap joint is as strong as the weakest member, even though the filler metal is low in strength. In a butt joint, the strength of the joint depends partly on the strength of the filler metal.

The basic joint design for adhesive bonding is the lap joint, and the type of loading is shear loading. The two types of parts generally encountered are flat parts and circular parts, with the shear areas as indicated in Fig. 15.1. The basic expression for calculations is:

$$\tau = P/A \tag{15.6}$$

A. Flat Parts

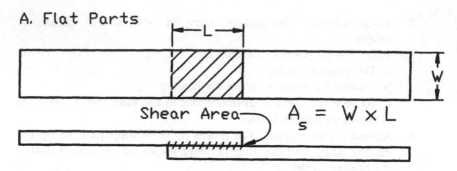

B. Circular (Tubing) Parts

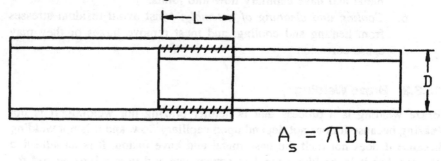

Figure 15.1 Shear areas for adhesive joints.

where

> τ = shear stress of adhesive
> A = total shear area
> P = load to be supported

This can be rewritten to solve for the shear area:

$$A = P/\tau \qquad\qquad\qquad\qquad (15.7)$$

Usually, one is trying to determine the length of the lap, which is dependent upon whether flat or circular parts are used. The expression for the length of lap for flat parts would be:

$$L = A/W \qquad\qquad\qquad\qquad (15.8)$$

where

> W = width of piece

The length of lap for circular parts, such as pipes, would be:

$$L = A/\pi D \qquad (15.9)$$

where

D = outside diameter of smaller pipe

Example Problem 15.4. A lap joint is made of two aluminum strips similar to the flat parts in Fig. 15.1A. Each aluminum strip has a tensile strength of 64,000 psi, and is 2 in. wide, $\frac{1}{8}$ in. thick, and 15 in. long. What would be the length of lap necessary to cause failure in the base metal, if the brazing filler metal has a strength of 12,000 psi? Use a safety factor of 2 in the design.

Solution. The first step is to determine the load that the straps can support in tension. This is the same load the joint will need to support in shear. The load that the straps can support is:

$$P = A \times \sigma$$

$$= (2 \text{ in.} \times \tfrac{1}{8} \text{ in.}) \times 64,000 \text{ psi}$$

$$= 16,000 \text{ lb}$$

Since the factor of safety is 2, the shear load design should be 32,000 lb. The shear area should be:

$$A = 32,000 \text{ lb}/12,000 \text{ psi}$$

$$= 2.667 \text{ in.}^2$$

But from Eq. (15.8), the length of lap is:

$$L = A/W$$

$$= 2.667 \text{ in.}^2/2 \text{ in.}$$

$$= 1.333 \text{ in.}$$

With the length of lap of 1.333 in., the strap would tend to fail before the joint fails, since the joint has a safety factor of 2.0 and the strap has a safety factor of 1.0.

15.4 COHESIVE (WELDING) JOINING PROCESSES

15.4.1 Introduction to Welding Processes

The cohesive, or welding, joining processes indicate a bonding between the two parts or the parts and filler metal, which is cohesive in nature. Usually,

Table 15.4 Welding Classification by Energy Source

Type of energy source and example processes	Energy level (W/m²)
Electrical	$10^5 - 19^8$
Gas resistance: SMAW, SAW, GMAW, GTAW	
Solid resistance: resistance spot, seam	
Liquid resistance: electroslag	
Mechanical	$10^4 - 10^6$
High pressure: friction, explosion	
Ultrasonic: for thin materials	
Chemical	$10^6 - 10^8$
Oxyfuel: acetylene: MAPP, propane	
Exothermic reaction: thermit	
Optical (beam, ray)	$10^{10} - 10^{12}$
Laser: CO_2, Ruby	
Electron beam: X-rays can be a problem	

high temperatures (at or near the melting point) are involved, but in a few instances the bond can be achieved using high forces. There are various methods for classifying welding process, but one of the more interesting approaches is by an energy source. There are four main energy sources for welding. These sources are presented in Table 15.4 with the approximate energy level.

In general, the higher the energy source, the deeper the penetration and the fewer the passes necessary to complete the weld. The electron beam and laser processes can weld more than 1-in. (25 mm) thick pieces in a single pass, whereas the gas resistance processes are limited to about $\frac{1}{4}$ in. (6 mm) in a single pass. The most common of the welding processes are the gas resistance processes, and most of this section will pertain to those processes.

15.4.2 Electrode Codes, Welding Positions, Welding Current Types, Welding Processes, and Welding Joints

The electrode codes for the welding electrodes used in the shielded metal arc welding (SMAW), or "stick electrode," process give considerable information about the requirements for welding. The code consists of a letter and four digits (for example, E-7014):

L-1234

where

L = letter code
1 = first digit
2 = second digit
3 = third digit
4 = fourth digit

The letter code is usually an "E," which means that it is a welding electrode. Occasionally an "R" may be used, which says that it is a rod for brazing or braze welding. The first two digits indicate the tensile strength of the filler metal. For example, for a code of E-7014, the 70 indicates that the tensile strength of the filler metal is 70,000 psi, or 70 kpsi (480 MPa). Most electrodes have tensile strength values of 60, 70, 80, 90, 100, and 110 kpsi (410, 480, 550, 620, 690, and 760 MPa, respectively). If the tensile strength is either 100 kpsi (690 MPa) or 110 kpsi (760 MPa), an extra digit is in the code for the third digit of the tensile strength.

The third digit of the code represents the welding position or welding positions for which the electrode can be used. There are four recognized welding positions: flat, horizontal, vertical, and overhead. The welding positions indicated by the third digit are:

Value of digit	Welding positions for electrode
1	All positions—flat, horizontal, vertical, and overhead
2	Flat and horizontal fillet welds only
3	Flat position only
4	Vertical down welds only

In practice, nearly all electrodes are coded either 1 or 2 in the third digit.

The fourth digit indicates the current to be used and the type of coating for the electrode. The coatings and currents used with the shielded steel electrodes are:

Digit	Coating type	Recommended current type
0	Cellulose, sodium	DCEP
1	Cellulose, potassium	AC or DC[a]
2	High titania, sodium	AC or DC

Coatings and currents continued

Digit	Coating type	Recommended current type
3	High titania, potassium	AC or DC
4	Iron powder, titania	AC or DC
5	Low hydrogen, sodium	DCEP
6	Low hydrogen, potassium	AC or DCEP
7	Iron powder, iron oxide	AC or DC
8	Iron powder, low hydrogen	AC or DCEP

ªDC implies either DCEP or DCEN can be used

In welding, there are three different types of welding currents that are used: AC and two types of DC. The welding currents can be classified as:

1. AC (alternating current)
2. DC (direct current)
   a. DCSP (direct current, straight polarity) or DCEN (direct current, electrode negative)
   b. DCRP (direct current, reverse polarity) or DCEP (direct current, electrode positive)

For processes with consumable electrodes, DCEP (DCRP) gives deeper penetration. The consumable electrode processes are those processes where the electrode is the filler metal. In processes with nonconsumable electrodes, such as the gas tungsten arc welding process, deeper penetration is achieved with the DCEN (DCSP) current. Most of the heat in welding is generated at the positive pole, so for consumable electrodes the heat is required to melt the electrode, whereas in the nonconsumable electrode processes one does not want to melt the electrode because it will contaminate the weld. Some of the more common arc welding processes and codes follow.

Abbreviation	Process name	Electrode type
SMAW	Shielded metal arc welding	Consumable
GTAW	Gas tungsten arc welding	Nonconsumable
GMAW	Gas metal arc welding	Consumable
SAW	Submerged arc welding	Consumable
PAW	Plasma arc welding	Nonconsumable
OAW	Oxyacetylene welding	—

A brief description illustrates the key features of the processes for some of the widely used arc welding processes (4).

15.4.2.1 Shielded Metal Arc Welding (SMAW)

SMAW is a basic arc welding process. Usually, the workpiece is grounded while the positive terminal is used to hold the electrode. This is also known as DCEP, or reverse polarity. An arc occurs between the consumable electrode and the workpiece. The electrode and the workpiece are melted and fused. In SMAW, the electrode is covered with a flux that burns and provides the shielding gas. Some of the flux coating solidifies as slag on the weld. This is a manual welding process, and is being replaced by other welding processes because of the relatively slow travel speeds and deposition rates. It has been used primarily for welding steels.

15.4.2.2 Gas Tungsten Arc Welding (GTAW)

GTAW uses a nonconsumable tungsten electrode and is shielded with an inert gas. Since it is undesirable to melt the electrode, DCEN (straight polarity) or AC current is used for most applications. An arc occurs between the tip of the electrode and the workpiece. The arc melts the workpiece and the filler metal to form a weld. The shielding gas is used to protect the arc and the surface from air. Therefore, no flux is needed to protect the surface. This process was formerly called "TIG." The shield gases commonly used are helium and argon. The process has been commonly used for welding aluminum and stainless steel materials.

15.4.2.3 Gas Metal Arc Welding (GMAW)

GMAW uses a shielding gas to protect the weld from air. A consumable wire electrode is fed automatically into the weld pool. This process was formerly called "MIG" welding. The shielding gas is usually helium, argon, or CO_2. This process was used primarily for welding ferrous materials, but new solid state power supplies have made it possible to weld aluminum via this process. This is a consumable electrode process, and DCEP (reverse polarity) and AC currents are commonly used.

15.4.2.4 Submerged Arc Welding (SAW)

In SAW, the flux consists of unfused granular particles. The flux is supplied through a tube in front of the path of the electrode. The continuously fed electrode is submerged in the flux, where an arc is created between the electrode and the workpiece. The arc is submerged in the flux, which protects the weld metal and the workpiece from the air. Since the arc is submerged, the eye shielding lens required is not as dark as the lens required for other arc welding processes. This process has been used primarily for welding steels. High travel speeds and high penetration depths are possible with this process, but it is somewhat limited to the flat and horizontal positions because the granular flux cover may fall off the weld materials in the vertical and overhead positions. Since SAW is a consumable electrode process, DCEP (reverse polarity) and AC currents are commonly used.

15.4.2.5 Plasma Arc Welding (PAW)

PAW has greater advantages than the previously mentioned welding processes. It has greater welding penetration and can be used to weld at high speeds. PAW has a nonconsumable electrode made of tungsten. Plasma gas is supplied over the tungsten electrode to form the plasma arc. Plasma gas is the same gas used in other welding processes, except the gas is ionized. PAW also uses a nonionized shielding gas to protect the welding zone from air while filler metal is being applied. The plasma gas provides high energy into the weld and increases the penetration. Since PAW is a nonconsumable electrode process, DCEN (straight polarity) and AC currents are typically used.

15.4.2.6 Oxyacetylene Arc Welding (OAW)

OAW uses heat produced by a chemical gas flame. OAW is a mixture of acetylene and oxygen. The flame is used for melting the base metal and the filler metal. The energy to form the flame is low compared to the arc welding processes, so the penetration depth and travel speeds are much lower. In addition to welding, the process is used for braze welding, brazing, and soldering.

There are five different types of weld joints and several different types of welds. The most common types of weld joints and weld types are:

Weld joints	Weld types
Butt weld	Fillet
Tee weld	Square
Corner weld	Bevel groove
Lap weld	V-groove
Edge weld	J-groove
	U-groove

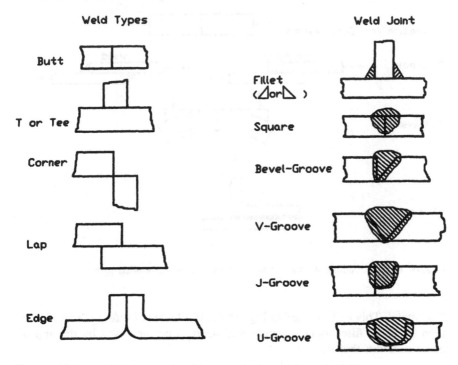

Figure 15.2 Weld joints and weld types. (Adapted from Ref. 6.)

Figure 15.2 illustrates the various weld joints and some of the weld grooves. The fillet weld is assumed to be an isosceles triangle, for this gives the largest shear area per unit volume of weld. There are four different types of loading stresses that must be of concern in welding and adhesive bonding (see Fig. 15.3):

1. Tension or compression
2. Shear
3. Bending
4. Peel (exaggerated cleavage)

15.4.3 Welding Design Calculations

The basic welding design calculations are for the tensile loading of butt welds and the shear loading of fillet welds. There are several excellent sources on welding design (5–8), and the materials presented here are a composite of these sources. The tensile loading of a butt weld requires only that the tensile strength of the weld filler metal be stronger than that of the

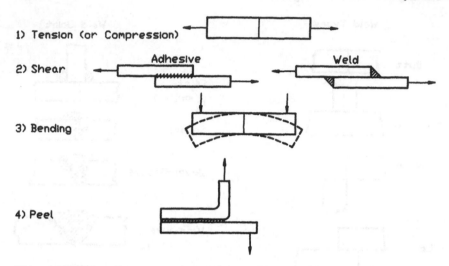

1) Tension (or Compression)

2) Shear

3) Bending

4) Peel

Figure 15.3 Loading stresses in welded and adhesive-bonded joints.

base metal. This is illustrated in Fig. 15.4a. If a factor of safety is considered, then the weld filler metal strength will need to be increased by the appropriate amount; that is:

$$\sigma_w \geq F_s \times \sigma_b \tag{15.10}$$

where

σ_w = tensile strength of weld filler metal
σ_b = tensile strength of the base metal
F_s = factor of safety for weld metal/joint design

The shear loading of fillet welds requires the determination of the shear strength. The shear failure is assumed to occur at 45°, and the weld length necessary to prevent failure can be calculated as a function of the weld size.

The first relationship needed is the relationship between the shear strength of the weld and the tensile strength of the weld material. That is, the shear strength used for calculations is 30 percent of the tensile strength of the weld metal, which is indicated by the first two digits of the electrode code.

$$\tau = 0.3\sigma_w \text{ (tensile strength of weld)} \tag{15.11}$$

where

τ = shear strength of weld
σ_w = tensile strength of weld filler metal

a) Tensile Loading - Butt Welds

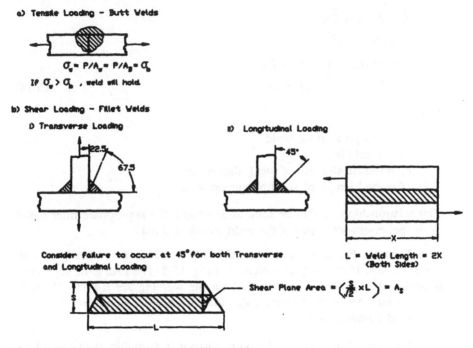

$\sigma_u = P/A_w = P/A_b = \sigma_b$

If $\sigma_u > \sigma_b$, weld will hold.

b) Shear Loading - Fillet Welds

D Transverse Loading   I) Longitudinal Loading

Consider failure to occur at 45° for both Transverse   L = Weld Length = 2X
and Longitudinal Loading   (Both Sides)

Shear Plane Area = $\left(\frac{s}{\sqrt{2}} \times L\right)$ = A_s

Figure 15.4 Loading in butt and fillet welds.

The shear area of the weld can be expressed as a function of the weld size and weld length, as indicated in Fig. 15.4:

$$A_s = s/2 \times L \qquad\qquad (15.12)$$

where

A_s = shear area
s = weld size
L = length of weld (total length if double fillet weld)

Since

$$\tau = P/A_s \qquad\qquad (15.13)$$

where

P = design load for weld

the weld length can be evaluated in terms of the weld size and weld metal properties via:

$$L = A_s \times \sqrt{2}/s$$
$$= P/\tau \times \sqrt{2}/s$$
$$= P/(0.3 \times \sigma_w) \times \sqrt{2}/s$$
$$= P/S \times \sqrt{2}/(0.3 \times \sigma_w) \qquad\qquad (15.14)$$

where

P = design load for weld
s = weld size
σ_w = tensile strength of weld filler metal
L = total length of weld to support load

This relationship can also be used to determine the appropriate filler metal or the necessary weld size if the weld length is fixed.

Example Problem 15.5. A double fillet weld is to be made on two pieces of steel 10 in. long, as indicated in Fig. 15.5. The load to be supported is 50,000 lb, and the electrodes used for the welding are E-8024. What is the weld size necessary to ensure that the load will be supported? Would a $\frac{3}{16}$-in. weld be adequate?

Solution. The total weld length, since it is a double weld, would be

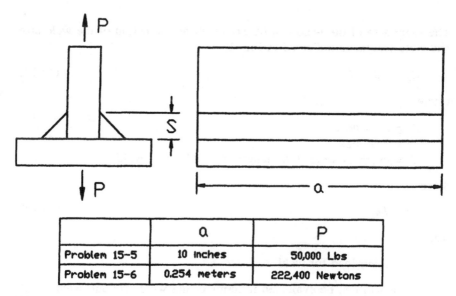

	a	P
Problem 15-5	10 inches	50,000 Lbs
Problem 15-6	0.254 meters	222,400 Newtons

Figure 15.5 Illustration for Example Problems 15.5 and 15.6.

20 in., that is, 10 in. on each side. Rearranging Eq. (15.14) to determine the weld size, the expression becomes:

$$s = P/L \times \sqrt{2}/(0.3 \times \sigma_w)$$
$$= (50{,}000 \text{ lb/20 in.}) \times \sqrt{2}/(0.30 \times 80{,}000 \text{ lb/in.}^2)$$
$$= 0.147 \text{ in.}$$

Since $\frac{3}{16}$ in. = 0.1875 in., a $\frac{3}{16}$-in. weld size would be more than adequate. Note that if an E-6024 electrode were used, the required weld size would be 0.196 in. and the $\frac{3}{16}$-in. weld would not be adequate.

Example Problem 15.6. A double fillet weld is to be made on two pieces of steel 0.254 m long, as indicated in Fig. 15.5. The load to be supported is 222,400 N, and the electrodes used for the welding are E-8024. What is the weld size necessary to ensure that the load will be supported? Would a 4.5-mm weld be adequate?

Solution. The total weld length, since it is a double weld, would be 508 mm, that is, 254 mm on each side. Rearranging Eq. (15.14) to determine the weld size, the expression becomes:

$$s = P/L \times \sqrt{2}/(0.3 \times \sigma_w)$$
$$= (222{,}400 \text{ N/508 mm}) \times \sqrt{2}/(0.30 \times 550 \text{ MPa})$$
$$= 3.75 \text{ mm}$$

A 4.5-mm weld size would be more than adequate. Note that if an E-6024 electrode were used, the required weld size would be 5.0 mm and the 4.5-mm weld would not be adequate.

15.4.4 Welding Defects

One of the problems of welding is that if improper materials or welding conditions are applied, catastrophic failure can result. In the past, several failures of welded structure have occurred that have made designers reluctant to use welds. Since high-strength materials have a tendency to be brittle, the failures can occur suddenly and be disastrous. Some of the past failures were embrittlement failures, which occurred at low temperatures, when steel incurred a ductile-to-brittle transition. These failures, such as the liberty ship failure, occurred when impact loads (such as waves during storms in cold weather) were applied. The causes of these failures are now understood, and the materials and welding procedures now used eliminate the occurrence of such failures.

There are various types of imperfections, sometimes called *defects*, that can occur during the welding process. The most common of the imperfections follow.

1. Cracks
   a. The worst type of imperfection
   b. No cracks are tolerated in the weld area
      i. Cracks that can occur during the welding process or during cooling:
         Cracks in the weld bead
         Cracks in the HAZ (heat-affected zone)
         Underbead cracking
         Lamellar cracking
         Hydrogen cracking
      ii. Incomplete fusion: weld metal is not hot enough to melt and fuse with base metal
      iii. Incomplete penetration: weld not at complete depth of pieces. (Incomplete penetration may be permitted in welds where the loading is low and strength is not a primary requirement.)
2. Porosity (gas holes)
   a. Some porosity may be permitted in welds
   b. Porosity can occur from not having clean surfaces, having moisture or oil on the surface, or from improper shielding
3. Inclusions (oxides)
   a. Some inclusions may be permitted in some types of welds
   b. Inclusions are caused by fluxes or oxides getting trapped in the weld metal and not coming to the surface
4. Welding problems
   a. Undercutting: melting base metal and not replacing with filler metal
   b. Splatter: spots of weld metal on base metal surfaces that may cause cracks in base metal or rough surfaces
   c. Excess convexity: excess metal deposited that can be costly
   d. Excess concavity: not enough metal for shear area

In some instances when a crack is found, a hole is drilled at the crack tip to reduce the stress concentration. This should be done only to prevent the weld from failing until it can be repaired.

The stress concentration for holes can be expressed in terms of the average stress, the major ellipse axis, and the radius of curvature at the edge of the hole; that is:

$$\sigma_{max} = \sigma_{ave} \times [1 + 2(c/r)] \tag{15.15}$$

where

σ_{max} = stress from concentration
σ_{ave} = average stress of part
c = major ellipse axis
r = radius of curvature

For a circular hole, the major ellipse axis is equal to the radius of the circle. The radius of curvature is also equal to the radius of the circle. So $c/r = 1$ for a drilled circular hole and the concentrated stress is equal to three times the average stress. This factor of 3 is the lowest value for a crack or hole, which is why holes are drilled at the tips of cracks to prevent their continued growth.

15.5 BASIC WELDING CALCULATIONS

There are several factors used to evaluate processes with respect to basic energy consumption and production parameters. Some of the more common expressions determine the energy per unit length of weld, the standard production time and production rate, and duty cycle considerations.

The energy per unit length is often used as a comparison base to evaluate different processes. It includes the three primary variables in welding: current, voltage, and travel speed. The expression is:

$$H = W/L \qquad\qquad (15.16)$$
$$= EI/S$$

where

W = energy
L = weld length
I = welding current (A)
E = welding voltage (V)
S = travel speed (in./min or mm/sec)
H = energy per unit length

The values are expressed in terms of J/in. or J/mm, and thus the travel speed must be converted to units of seconds, since a joule has units of V-A-sec. Thus a welding process with a current of 100 A, a welding voltage of 20 V, and a travel speed of 10 in./min would have an energy per unit length of 12,000 J/in.

The three basic variables in the welding process are current, voltage, and travel speed. The current is the primary variable used to control the depth of penetration. The weld voltage is related to the arc length and affects the width of the weld bead. The travel speed relates to the movement of the electrode and deposition of the weld bead. It also affects the weld bead shape, and it is used to control the energy input into the weld to ensure proper melting, since the current and voltage tend to be kept constant during the welding process.

The time to produce a weld is given by the following expression:

$$T = L/(S \times OF) \tag{15.17}$$

where

T = time for weld (min or sec)
L = length of weld (in. or mm)
S = travel speed (in./min or mm/sec)
OF = operator factor

The operator factor is expressed as the ratio of the arc time to the total weld cycle time:

$$OF = arc\ time/total\ cycle\ time \tag{15.18}$$

where

arc time = time welding machine is arcing
total cycle time = arc time plus preparation time (positioning work-pieces, etc.) and cleanup time (removing slag, etc.)

The production rate can be determined from the production time; that is:

$$P = 60/T \tag{15.19}$$

where

T = production time, in min/piece
P = production rate in pieces/hr

The operator factor can vary considerably for the different welding processes and types of work. For the manual shielded metal arc welding process, the operator factor will typically vary from 0.10 to 0.40, whereas for automatic welding processes the operator factor will vary from 0.40 to 0.90, and for semiautomatic processes, the range would be from 0.20 to 0.60.

Multiple-pass welding is when several passes are required to complete the weld. The passes may be at different travel speeds; and rather than calculate the time for each pass, an overall travel speed can be used in the formulas. The expression for the overall travel speed is:

$$S = 1/(1/S_1 + 1/S_2 + 1/S_3 + \cdots + 1/S_n) \tag{15.20}$$

where

S = overall travel speed
S_1 = travel speed for 1st pass
S_2 = travel speed for 2nd pass
S_n = travel speed for last pass

The duty cycle of the power supply relates the percentage of time of a 10-min interval that the power supply can operate at the rated voltage and current. There are usually three different duty cycles for machines: 20 percent for hobby and craft machines, 60 percent for industrial manual and semiautomatic processes, and 100 percent for industrial automatic processes. The relationship to evaluate the actual welding current is:

$$I_a^2 \times D_a \le I_R^2 \times D_R = \text{constant} \qquad (15.21)$$

where

I_a = actual welding current
D_a = actual duty cycle
I_R = rated welding current of power supply
D_R = rated duty cycle of power supply

Example Problem 15.7. If a power supply has a duty cycle of 60 percent and rated current of 200 A, can the power supply operate continuously at 130 A?

Solution. Eq. (15.21) to solve for the actual duty cycle; if it is greater than 100 percent, then the power supply can operate continuously:

$$D_a \le (I_R/I_a)^2 \times D_R$$
$$\le (200/130)^2 \times 60$$
$$\le 142 \text{ percent (which is greater than 100 percent)}$$

This implies that the power supply can operate continuously at 130 Amps, for the actual duty cycle is more than 100 percent when the current is only 130 A. Note that if the current were 160 A, the power supply would overheat, because the actual duty cycle would be less than 100 percent (approximately 94 percent).

The area of the ideal fillet weld is assumed to be an isosceles triangle, but the weld is actually a convex or concave weld. The adjustment to the ideal area for a convex weld is 1.21 and for a concave weld is 1.15. That is, a convex weld has 1.21 times the weld cross-sectional area of the ideal fillet weld. The adjustment factors are often used in calculating the weld metal requirements. Figure 15.6 illustrates the ideal, convex, and concave fillet welds.

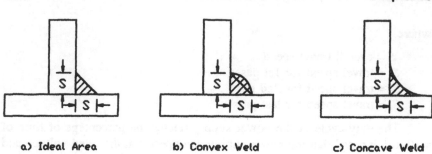

a) Ideal Area   b) Convex Weld   c) Concave Weld

Figure 15.6 Ideal, convex, and concave weld areas for fillet welds of equivalent weld size "*s*."

15.6 WELDING ECONOMICS

The economics of the welding process is one of the major criteria in selecting the particular process. The welding economic evaluation procedures are consistent among authors (5,7,9) and generally consist of determining the fixed and variable cost elements. These cost elements usually consist of the following components:

Fixed costs	Variable costs ($/ft or $/lb)
Capital investment (machine)	Labor
Tooling and fixturing	Overhead
	Material (electrode)
	Power
	Shielding
	a. Gas
	b. Flux

The capital investment cost for the machine is depreciated over the life of the investment and will depend upon the depreciation method utilized. The special tooling and fixturing costs are usually expensed on that order. The emphasis will therefore be placed upon the determination of the variable costs.

The basis of the variable costs can be either $/ft or $/lb, and the base selected will be $/ft, because this can usually be easily obtained from the drawings. The values can be easily converted to $/lb values by:

$$\text{Cost (\$/lb)} = \text{Cost (\$/ft)} \times 1 \text{ ft/12 in.} \times [1/\text{weld area (in.}^2)]$$
$$\times [1/\text{metal density (lb/in.}^3)] \qquad (15.22)$$

where

weld area = weld cross-sectional area per unit length
$$= \tfrac{1}{2} \times s^2 \text{ for fillet welds}$$

Expressions for each of the five variable-cost items are developed in the units of \$/ft of weld length.

1. *Variable labor cost (VLC)*:

$$\text{VLC (\$/ft)} = [\text{LC (\$/hr)} \times 12 \text{ in./ft}]/[S \text{ (in./min)}$$
$$\times 60 \text{ min/hr} \times \text{OF}] \qquad (15.23)$$

where

VLC = variable labor cost (\$/ft)
LC = labor rate (\$/hr)
S = travel speed (in./min)
OF = operator factor

2. *Variable overhead cost (VOC)*:

$$\text{VOC (\$/ft)} = [\text{OH (\$/hr)} \times 12 \text{ in./ft}]/[S \text{ (in./min)}$$
$$\times 60 \text{ min/hr} \times \text{OF}] \qquad (15.24)$$

where

VOC = variable overhead cost (\$/ft)
OH = overhead rate (\$/hr)
S = travel speed (in./min)
OF = operator factor

3. *Variable material (electrode) cost (\$/ft)*: There are two different cases to consider for the variable material costs: (1) the case where the welding is performed manually, and (2) the case where the welding is done with automatic wire feed. The material cost expression for automatic welding is:

$$\text{VEC}_A = [\text{WF (in./min)} \times \text{EWL (lb/in.)} \times \text{EC (\$/lb)}$$
$$\times 12 \text{ in./ft}]/[S \text{ (in./min)} \times \text{EML}] \qquad (15.25)$$

where

VEC_A = variable material cost for automatic welding (\$/ft)
WF = wire feed rate (in./min)
EWL = wire weight per length of wire (lb/in.)

EC = cost of electrode ($/lb)
S = travel speed (in./min)
EML = electrode metal yield

The electrode metal yield is determined by:

EML = (weld metal deposited)/(electrode material consumed)   (15.26)

For automatic and semiautomatic welding processes, the value of the electrode metal yield is from 0.80 to 1.00. For the manual welding process, such as SMAW, the electrode metal yield is approximately 0.60, because the electrode coating and electrode stubs greatly reduce the yield.

The material cost expression for manual welding is:

$$VEC_M = [PWD\ (lb/ft) \times EC\ (\$/lb)]/EML \qquad (15.27)$$

where

VEC_M = variable material cost for manual welding ($/ft)
PWD = pounds of weld metal deposited per length of weld (lb/ft)
EC = cost of electrodes ($/lb)
EML = electrode metal yield

PWD, the pounds of weld metal deposited, can be determined from the weld cross-sectional area and metal density:

$$PWD = A_w\ (in.^2) \times density\ (lb/in.^3) \times 12\ in./ft \qquad (15.28)$$

where

A_w = weld cross-sectional area per unit length of weld

The weld area may be adjusted for a convex or concave weld.

4. *Variable power cost*:

$$VPC\ (\$/ft) = [(I \times E)/1000 \times PC\ (\$/kW\text{-}hr)$$
$$\times 12\ in./ft]/[S\ (in./min) \times 60\ min/hr \times M] \qquad (15.29)$$

where

I = welding current (A)
E = welding voltage (V)
PC = power cost ($/kW-hr)

5. *Variable shielding cost*: The shielding used for welding is usually either a shielding gas, as in the GTAW and GMAW processes, or a flux, as in the submerged arc welding (SAW) process. The expression for the variable gas cost of the shielding gas is:

$$\text{VGC (\$/ft)} = [\text{GFR (ft}^3\text{/hr} \times \text{GC (\$/ft}^3)$$
$$\times \ 12 \text{ in./ft]/[}S \text{ (in./min)} \times 60 \text{ min/hr]} \qquad (15.30)$$

where

GFR = gas flow rate (ft^3/hr)
GC = gas cost (\$/ft^3)
S = travel speed (in./min)

The expression for the variable flux cost is:

$$\text{VFC (\$/ft)} = [\text{PWD (lb metal/ft)}$$
$$\times \text{ FCR (lb flux/lb metal)} \times \text{FC (\$/lb)]} \qquad (15.31)$$

where

VFC = variable flux cost (\$/ft)
PWD = pounds of weld metal deposited per unit of weld (lb/ft)
FCR = flux consumption rate (lb flux/lb metal deposited)
FC = flux cost (\$/lb)

The total variable cost is the sum of the appropriate variable cost expressions for that process. Generally, 80–90 percent of the costs are labor, overhead, and material costs. The power and shielding costs tend to be small compared to the other costs. The corresponding formulas with metric units are found in Creese, Adithan, and Pabla (9).

Example Problem 15.8. A weld 20 in. in length was to be made using the submerged arc welding process. Determine the variable cost per foot and the number of pieces produced per hour. The setup time is 20 min, and a lot of 20 parts is to be made. The process data is in Table 15.5

Solution. The variable labor cost is:

$$\text{VLC (\$/ft)} = (\$10\text{/hr} \times 12 \text{ in./ft})/(60 \text{ in./min} \times 60 \text{ min/hr} \times 0.50)$$
$$= 0.0666\text{/ft}$$

$$\text{VOC (\$/ft)} = (\$30\text{/hr} \times 12 \text{ in./ft})/(60 \text{ in./min} \times 60 \text{ min/hr} \times 0.50)$$
$$= \$0.200\text{/ft}$$

$$\text{VEC}_a \text{ (\$/ft)} = (80 \text{ in./min} \times 0.0035 \text{ lb/in.} \times \$0.25\text{/lb} \times 12 \text{ in./ft})/(60$$
$$\text{in./min} \times 0.95)$$
$$= \$0.0147\text{/ft}$$

$$\text{VFC (\$/ft)} = (0.00708 \text{ lb/in.} \times 0.20 \text{ lb flux/lb metal} \times \$0.15\text{/lb flux}$$
$$\times \ 12 \text{ in./ft})$$
$$= 0.00255\text{/ft}$$

Table 15.5 Data for Example Problem 15.8

Process data

Process	SAW (submerged arc welding)
Travel speed	60 in./min
Welding current	520 A
Welding voltage	26 V
Machine efficiency	0.95 (95%)
Duty cycle	100%
Operator factor	0.50
Electrode metal yield	0.99
Gas flow rate	Not applicable
Wire feed rate	80 in./min
Electrode wt/length	0.0035 lb/in.
Flux consumption rate	0.2 lb flux/lb metal
Deposition rate	Calculate from travel speed
Weight of weld metal deposited	0.00708 lb/in.

Cost data

Labor rate	$10/hr
Overhead rate	$30/hr
Gas cost	Not applicable
Flux cost	$0.15/lb
Electrode cost	$0.25/lb
Power cost	$0.05/kW-hr

$$\text{VPC (\$/ft)} = \frac{(520 \text{ A} \times 26 \text{ V})/1000 \times \$0.05/\text{kW-hr} \times 12 \text{ in./ft}}{60 \text{ in./min} \times 60 \text{ min/hr} \times 0.95}$$
$$= \$0.00238/\text{ft}$$

Total variable cost = 0.0666 + 0.200 + 0.0147 + 0.00255 + 0.00238
= $0.286/ft

The production rate is determined by:

T = 20 in./(60 in./min × 0.50)
  = 0.667 min

P = 60 min/hr/0.667 min/piece
  = 90 pieces/hr

The time to produce the lot of 20 pieces would be:

T (lot) = 20 min (setup) + 20 pieces × 0.667 min/pc

  = 20 min + 13.34 min

  = 33.34 min

The total cost for the lot of 20 pieces would be:

TC (lot) = 20 min (setup)/60 min/hr × ($10/hr + $30/hr) + 20 pieces × 20 in./part × 1 ft/12 in. × $0.286/ft

  = $13.33 + $9.53

  = $22.86

15.7 SUMMARY AND CONCLUSIONS

The joining processes can be considered in three major groups: fasteners, adhesive processes, and cohesive processes. The fasteners group includes those for nonpermanent assembly as well as those for permanent assembly. The adhesive processes include the organic adhesives, such as glues and epoxies, and inorganic adhesives, such as soldering and brazing. The cohesive processes include the various electrical welding, chemical welding, mechanical welding, and optical beam welding processes. The major welding defects were described, with the most serious being cracks.

Basic design considerations were considered for bolts, lap adhesive joints, and welded joints. The economic considerations were evaluated for the arc welding processes, and a detailed calculation procedure for arc welding processes was presented. The major cost items for arc welding generally are the labor and overhead costs, which promotes investment in automation to reduce welding times.

15.8 EVALUATIVE QUESTIONS

1. What are the three major categories of joining processes?

2. List the two major categories of fasteners, and give an example of each.

3. What are the three types of failure for a threaded fastener, and which one is used for design?

4. What should be the minimum length of engagement of a $\frac{1}{2}$-13, grade 1 bolt with a cast iron block.

Bolt material		Nonbolt Material		
Gr 5 UTS	120,000 psi	Cast iron	UTS	27,000
Gr 5 USS	92,000 psi	Cast iron	USS	33,500
Gr 1 UTS	60,000 psi	PM aluminum	UTS	23,500
Gr 1 USS	36,000 psi	PM aluminum	UTS	13,900
		SAE 903 zinc	UTS	41,000
		SAE 903 zinc	USS	31,000

Bolt size	A (tension) (in.^{2})	A (shear-bolt) (in.^{2}/thread)	A (shear-hole) (in.^{2}/thread)
$\frac{1}{2}$-13	0.1419	0.0599	0.0866
$\frac{1}{4}$-20	0.0318	0.01840	0.02699

a. What do the $\frac{1}{2}$ and 13 represent?

b. What will fail in shear—bolt or block? (bolt, but design is such that the bolt should fail in tension rather than shear.)

c. Determine the length of bolt engagement. (0.304 in.)

5. What are the five steps in the application of an adhesive?

6. A lap joint is to be made connecting a $\frac{1}{8}$-in.-thick strip of stainless steel to a $\frac{1}{4}$-in.-thick strip of aluminum. The strips are 1 in. wide, and the epoxy used has a shear strength of 3000 psi; the aluminum has a tensile strength of 20,000 psi, and the stainless steel has a strength of 80,000 psi. If a safety factor of 3 is applied, over what area should the epoxy be applied?

a. Make a sketch of the joint.

b. Calculate the design load. (15,000 lb)

c. Calculate the shear area. (5 in.^{2}, or L = 5 in.)

d. What do you expect to fail? Explain your answer. (Aluminum)

7. What are the primary differences between brazing and gluing? between brazing and welding? between brazing and braze welding? between brazing and soldering?

8. What is the six-step procedure for brazing or soldering?

9. A piece of copper pipe is soldered to another pipe with a fitting. The copper pipe has an external diameter of $\frac{1}{2}$ in., the thickness is 0.050 in., and the fitting has an internal diameter of $\frac{1}{2}$ in. plus 0.015 in. If the copper pipe has a tensile strength of 50,000 psi and the solder has a shear strength of

4000 psi, what would be the length of lap to prevent failure in the soldered joint? (0.56 in.)

10. What are the four types of loading stresses that must be considered in welding and adhesive bonding?

11. What are the four main types of energy sources used in welding?

12. A welding electrode is coded E 6024.
    a. What is the tensile strength of the electrode material?
    b. In what welding positions can the electrode be used?
    c. What is the type of electrode coating?
    d. What type of current(s) can be used for the electrode?

13. a. What are the two different types of direct current?
    b. Which current type gives the best penetration for SMAW?
    c. Which current type gives the best penetration for GTAW?
    d. Which current type gives the best penetration for SAW?

14. What are the five types of weld joints?

15. What are the four main groups of weld imperfections, and which type is the worst?

16. What are the three main process variables in the gas arc welding processes?

17. Two pieces of steel, 6 in. long, 2 in. wide, and $\frac{1}{8}$ in. in thickness, are butt welded together, with the long sides joined together. The power supply is rated at 200 A with a 20 percent duty cycle. The electrodes are E 7014, $\frac{1}{8}$ in. in diameter, and the welding conditions are 160 A and 24 V. The travel speed is 10 in./min, and the operator factor is 0.60. Assume a production run of 1000 welds is to be made.
    a. What is the weld time per piece? (1 min/piece)
    b. What is the arc time per piece? (0.60 min/piece)
    c. What is the production rate? (60 pcs/hr)
    d. What is the energy per unit length? (23,040 J/in.)
    e. Will the power supply overheat? Explain your answer with numerical calculations. (yes)

18. A 7014 electrode was used to weld two $\frac{1}{4}$-in. steel strips as indicated in Fig. 15.7. The steel has a tensile strength of 60,000 psi, the welds are full depth ($\frac{1}{4}$ in.), and a safety factor of 2 is used. The larger width is 2 in. and the small width is 1 in. Determine the weld length L in Fig. 15.7. (4 in.)

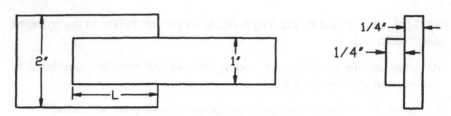

Figure 15.7 Lap weld dimensions to determine lap length for safe weld.

19. Two pieces, 12 in. long, are welded together with 7014 electrodes. One piece, 3 in. wide and $\frac{1}{2}$ in. thick, is fillet welded to the second piece, 6 in. wide and $\frac{1}{2}$ in. thick. The 3-in. piece is positioned on the center of the 6-in. piece and fillet welded on both sides to the top side of the 6-in.-wide piece. The welds are full length, or 12 in. The power supply has a 60 percent duty cycle and is rated at 200 A. The travel speed is 15 in./min, the operator factor is 0.80, the welding current is 150 A, and the welding voltage is 20 V. The density of the steel is 0.28 lb/in.^{3} The operator rate is \$18.00/hr and the machine rate is \$24.00/hr. The weld size is $\frac{1}{4}$ in.

   a. What is the arc time to weld the two pieces together? (1.6 min)
   b. What is the allowed time to weld the two pieces together? (2.0 min)
   c. What is the energy per unit length? (12,000 J/in.)
   d. What is the weight of metal deposited for the two convex fillet welds if the extra percentage of metal above the theoretical amount is 20 percent? (0.252 lb)
   e. What is the deposition rate during welding? (0.1575 lb/min)
   f. What is the total weld shear area for design calculations? (4.24 in.^{2})
   g. What is the operator and machine cost for the weld (\$1.40)

20. A metal box is to be welded together. The base is 6 in. × 10 in., and the wall height is 4 in. All inside corners are to be $\frac{1}{4}$-in. fillet welds (no extra metal). The material is a low-carbon sheet steel 0.20 in. thick. Determine the time and cost to weld a lot of 40 boxes when the setup time is 24 min and the following data apply:

Process	Gas metal arc welding
Travel speed	16 in./min
Welding current	400 A
Welding voltage	27 V
Machine efficiency	0.90

Data continued

Process	Gas metal arc welding
Duty cycle	60%
Operator factor	0.50
Electrode metal yield	0.95
Labor rate	$12.00/hr
Overhead	$24.00/hr
Gas cost	$.30/ft^3
Electrode cost	$0.25/lb
Power cost	$0.10/kW-hr
Metal density	0.28 lb/in.^{3}
Gas flow rate	40 ft^3/hr
Wire feed rate	72 in./min
Electrode wt/length	0.00194 lb/in.
Deposition rate	Calculate
Wt of weld metal deposited	Calculate

a. What is the weight of the weld metal deposited? (0.00875 lb/in.)
b. What is the deposition rate? (8.4 lb/hr during welding)
c. What is the total length of weld? (4 ft)
d. What is the labor cost per ft? ($0.300/ft)
e. What is the overhead cost per ft? ($0.600/ft)
f. What is the material cost (electrode) per foot? ($0.0275/ft)
g. What is the gas cost per foot? ($0.150/ft)
h. What is the power cost per foot? ($0.015/ft)
i. What is the total cost for 40 units? ($189.20)
j. What is the total time for 40 units? (264 min, or 4.4 hr)

REFERENCES

1. Parmley, Robert O. Standard Handbook of Fastening and Joining, McGraw-Hill, New York, 1977.
2. Husen, R. A. Fastener Engineering Handbook, Volume 3: Design Data, 1979, Fastener Engineering Department of Ford Motor Company, Dearborn, MI.
3. Connor, L. P., ed. Welding Handbook—Welding Technology, 8th ed., Vol. 1, American Welding Society, 1989.
4. Connor, L. P., ed. Welding Handbook—Welding Technology, 8th ed., Vol. 2, American Welding Society, 1991. (Also other volumes for specific process details.)
5. Lindberg, R. A., and Braton, N. R. Welding and Other Joining Processes, Allyn and Bacon, Boston, 1976.

6. Blodgett, O. W. "Joint Design and Preparation," American Society for Metals, Metals Handbook, Volume 6: Welding, Brazing, and Soldering, 9th ed., 1983, Metals Park, OH, pp. 60–72.
7. Cary, H. B. Modern Welding Technology, Prentice Hall, Englewood Cliffs, NJ, 1989.
8. The Procedure Handbook of Arc Welding, 12th ed., 1973. Lincoln Electric Company, Cleveland, OH.
9. Creese, R. C., Adithan, M., and Pabla, B. S. Estimating and Costing for the Metal Manufacturing Industries, Marcel Dekker, New York, 1992, pp. 149–176.

INTERNET SOURCES

General history of welding process: *http://www.sybersteel.com/pages/history.htm*

Penton Publishing—chapters on fastening and joining from machine design: *http://www.penton.com/md/bde/rvfjtoc.html*

Adhesive bonding section gives good overview on methods and designs of adhesives: *http://www.eastman.com/ppbo/design/scontents.shtml*

Job knowledge for welders, including welding equipment, weldability of materials, and welding processes: *http://www.twi.co.uk/bestprac/jobknol/jobknol.html*

IV
MANUFACTURING SUPPORT FUNCTIONS

16
Tool Design for Manufacturing

16.1 INTRODUCTION

Tool design is the key to success for the manufacture of a component at a competitive price in all manufacturing processes. Tooling consists of a vast array of cutting tools, dies for casting, sheet metal working, and forging, gages, jigs and fixtures, and many other devices for forming or holding a part while it is being formed. Tool design is involved in all of the manufacturing processes, but the tools are quite different in shape and purpose for the different manufacturing processes. The tools may be special purpose, such as the pattern to produce a casting, or general purpose, such as a clamp to hold a part to table on a milling machine.

Tool designers must work with process planning engineers and product design engineers to ensure that products are produced at a competitive price with short lead times and with high quality. This requires design and graphic skills, understanding of process operations and production rates, and cost analysis capabilities.

The most common types of tooling (1) are:

1. Cutting tools, including single-point cutting tools, drills, reamers, milling cutters, broaches, taps, and saw blades.
2. Jigs and fixtures for guiding the tool and holding the workpiece. These devices can be for guiding arc welds as well as for holding devices for machining.
3. Dies for plastic modeling, die casting, permanent molding, wax pattern molds for investment casting, forging, core boxes, extrusion, and wire drawing.
4. Patterns for sand castings.
5. Cutting and forming tools for sheet metal work.
6. Gages and measuring instruments such as GO and NOT-GO plug-and-snap gages, indicating gages, pneumatic gages, electronic indicators, optical projectors, laser gages, and other devices.

Tool engineers are specialists in only one or two of the tooling types mentioned. The tool engineer is not only a designer of the tools, but must ensure that the tooling can be produced in a relatively short time. Tooling lead time is the major delay in most manufacturing processes, because most tools are specialized for a particular product. In the machining processes, tooling is relatively flexible, and this is one of the primary reasons why machining is such a dominant forming process. Customers want short lead times and will consider long lead times only if a significant cost advantage can be obtained by that manufacturing process. Since design is a major component of tool engineering, many tool engineers have backgrounds in manufacturing, production, or mechanical engineering.

This chapter is only an introduction to tool design and many references (1–5) have more detailed information on tool design. Tool engineering has not been regarded as an important area in most engineering programs in the United States, and this may be one of the reasons for the decline of its manufacturing base. Most engineering programs in the United States have deleted tool design and tool engineering courses during the past 30 years as pressure has been applied to reduce the credit requirements in engineering programs.

16.2 TOOL DESIGN FOR MACHINING

The purpose of tool design for machining is to make tools that will cut away or remove material to produce the desired shape. The tool design for machining can be separated into two areas: tool design for single-point cutting and tool design for multiple-point cutting (3). Multiple-point cutting generally has higher metal/material removal rates and can produce the desired shape faster, but it is usually more expensive.

The design of tools for single-point cutting begins with tool angles, cutting speeds, feeds, depth of cut, and the material being cut. Some of these factors were mentioned in Chapter 14 and will be repeated here. One of the primary considerations is that the cutting tool material be harder than the material being cut, which is one reason why tool steels, carbides, and ceramics are used. The tool geometry of the single-point tool is shown in Fig. 14.1. The side rake and back rake angles are 5–10°, and the end relief angle (also called the *front clearance angle*) is 7–10° (6) for tool steels cutting stainless steels. However, when carbide inserts are used, the rake angle (side) is 5–10° for small depth of cuts, but may be a negative 5° for large depth of cuts (7). Single-point cutting tools are generally used in turning operations.

Other factors, such as the material being cut and the use of cutting fluids, have an impact upon the tools being used. New tool materials are continually being developed to give longer tool lives and permit higher cutting speeds. Not only are there different tool materials, but coatings on the tool materials are used to improve greatly the performance of the tool material.

Multiple-point tools can be either a combination of single-point cutting tools in a cluster or tools with multiple cutting edges. The combination of single-point cutting tools in a cluster can permit multiple roughing cuts in the same pass. The use of both roughing and finishing cuts in the same pass is difficult, because one usually uses smaller feeds and higher speeds for finishing cuts than for roughing cuts.

The traditional multiple-point cutters are milling cutters, drill bits, reamers, and specialized tools for countersinking, spotfacing, counterboring, and tapping. The multiple-point cutters are basically a series of single-point tools mounted in such a manner that all teeth follow essentially the same path across the workpiece (3). The tool may travel in a rotary path, such as drills, taps, and milling cutters, or it may travel in a linear fashion, such as with broaches, hacksaws, and band saws.

One of the important factors for milling cutters is the determination of the number of teeth in the cutter. One expression (3) for the number of teeth in a high-speed steel cutter is:

$$n = 19.5\sqrt{R} - 5.8 \tag{16.1}$$

where

R = cutter radius (in.)

For example, if the diameter of the cutter were 5 in. (radius 2.5 in.), the number of teeth would be:

$$n = 19.5\sqrt{2.5} - 5.8$$
$$= 25.03$$
$$= 25 \text{ teeth}$$

For face milling with carbide cutters, the expression for the number of teeth is:

$$n = 1.6 \times D \tag{16.2}$$

where

D = cutter diameter (in.)

When the number of teeth is greater than 10, the number of teeth is an even number. Thus, if the cutter diameter were 10.5 in.,

$$n = 1.6 \times 10.5$$
$$= 16.8$$

The number of teeth would be 16 and not 17, since 16 is an even number.

The tool geometry and tool angles for multiple-point cutters are beyond the scope of this section and the references should be consulted. The number of teeth for a cutter is required for determining machine power requirements and for production time values.

16.3 JIGS AND FIXTURES

Jigs and fixtures are used to hold the workpiece in an exact position. A fixture is always fastened to a machine or bench in a fixed position, hence the name fixture. A jig is not fastened to the machine on which it is being used and may be moved around; it is meant to ensure that a hole will be machined in the proper location and is used exclusively for hole-making and -finishing operations such as drilling, tapping, reaming, and boring. Thus, if the operation includes milling, turning, planing, shaping, or the like, the device is a fixture and not a jig.

The purpose of jigs and fixtures is to prevent objects from moving, and this involves prevention or control of motion in the twelve directional movements, or degrees of freedom. Figure 16.1 shows the twelve degrees of freedom. The six axial degrees of freedom allow movement in both directions (plus and minus) along the three axes shown as X, Y, and Z. The six radial degrees of freedom allow rotational movement in both clockwise and counterclockwise directions about the X, Y, and Z axes. The twelve directional movements require a minimum of six contact points to prevent motion.

Locators are used for restricting workpiece movement and for resisting the cutting forces. Clamps are not as strong as locators and therefore should not be used for holding the workpiece against the cutting forces. Plane, concentric, and radial locators are three common types of locators used to restrict movement; the emphasis here will be on plane locators. Locators are sometimes used to support the part as well as for providing location. In these instances, the locators are often called *supports*.

The six-point, or 3-2-1, location principle provides a conceptual location method. According to the 3-2-1 location rule, a workpiece can be located completely by banking it against three points in one plane, two points in a second perpendicular plane, and one point in a third perpendicular plane, as illustrated in Fig. 16.2. The three points in one plane restrict five

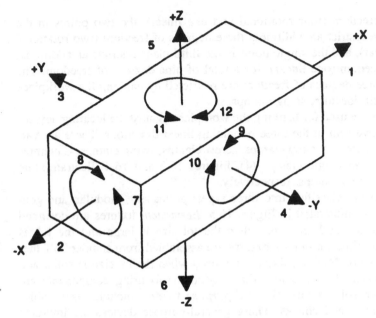

Figure 16.1 The twelve degrees of freedom in fixturing. (Courtesy of Carr Lane.)

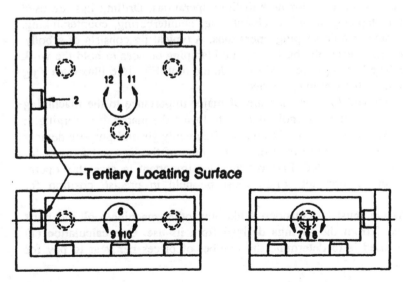

Figure 16.2 The 3-2-1 location method for restricting nine degrees of freedom. (Courtesy of Carr Lane.)

degrees of freedom (four rotational and one linear), the two points in the second plane restrict an additional three degrees of freedom (two rotational and one linear), and the single point in the third plane restricts an additional degree of freedom (one linear), for a total of nine degrees of freedom. The remaining three degrees of freedom are restricted by banking the workpiece against the six locators, using clamps.

Clamps are used for banking the workpiece against the locating devices to restrict movement in the three remaining linear directions. There are various groups of clamps: strap clamps, screw clamps, swing clamps, C clamps, cam clamps, and toggle clamps (4). Figures 16.3 and 16.4 illustrate strap clamps and toggle clamps, respectively.

Fixtures are classified into three types: permanent, modular, and general purpose, as illustrated in Figure 16.5. *Permanent* fixtures are designed for a specific job and are used when the lot size is large and the job is repeated many times. *Modular* fixtures are assembled from standard off-the-shelf components. Modular fixtures are used when the lot size is small and the job is repeated only a few times. Modular fixturing components are assembled and sold as kits. *General-purpose* fixtures include vises, table-mounted clamps, and chucks. These general-purpose devices are low-cost items and are used for simple jobs; however, they are often inadequate for complex parts or high-volume production.

A jig *guides* the cutting tool, whereas a fixture *references* the cutting tool. Most jigs are used for hole-drilling operations. Drilling jigs are used for drilling, tapping, reaming, chamfering, counterboring, countersinking, and other similar hole-shaping operations. A typical jig consists of a body, a jig plate that carries the bushings, and clamping devices to hold the work in the jig. The bushings are used to guide the tool; Fig. 16.6 illustrates a jig using a vise as the clamping device.

The jigs and fixtures used are of major importance to the machining operations. In casting, the mold is used to locate the material; in forging, it is the die. Although jigs and fixtures are frequently general-purpose devices and can be used for more than one part, some of the jigs and fixtures for complex parts are limited. Fixtures are also important in assembly operations, for the components often must be held in precise position for assembly.

A fixture must be considered like any other machine tool and must pay for itself from the savings derived from its use. One calculation (1) commonly used is to determine the number of pieces per year to pay for the fixture:

$$N = (Cf + Sr)/s(1 + L) \tag{16.3}$$

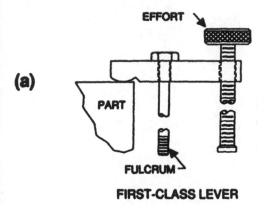

(a)

FIRST-CLASS LEVER

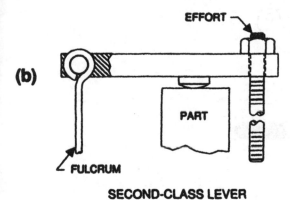

(b)

SECOND-CLASS LEVER

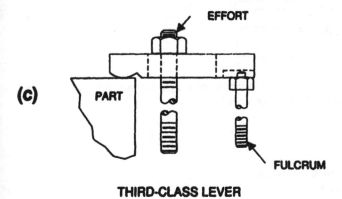

(c)

THIRD-CLASS LEVER

Figure 16.3 Strap clamps as first-, second-, and third-class levers. (Courtesy of Carr Lane.)

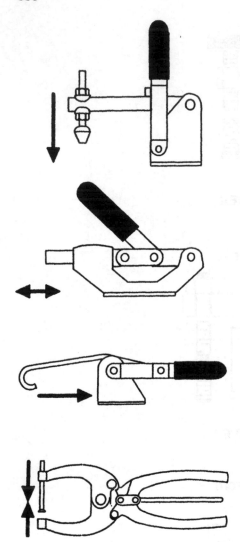

Figure 16.4 The four basic toggle actions for toggle clamps: hold down, pull/push, latch, and squeeze. (Courtesy of Carr Lane.)

where

  N = number of pieces per year
  C = fixture cost, \$
  f = fixture cost rate (return rate + operation cost rate + fixture overhead rate + depreciation rate) as decimal

Permanent Fixturing
(special purpose)

Modular Fixturing

General Purpose
(Vises, chucks,
subplates, and
table-mounted
clamps)

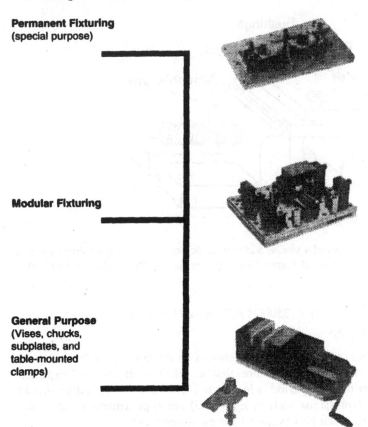

Figure 16.5 The three classes of fixtures: permanent fixturing, modular fixturing, and general-purpose fixturing. (Courtesy of Carr Lane.)

r = required rate of return
S = salvage value of fixture at end of 1st year, $
s = savings per piece from use of fixture, $/pc
L = total overhead rate (decimal)

Example Problem 16.1. A fixture has an estimated cost of $4000, the fixture cost rate is 0.60, the end of year value is expected to be $2500, the savings is $0.45 per piece, the required rate of return is 25 percent, the fixture life is 3 yr, and the total manufacturing overhead rate is 40 percent. What is the number of pieces that must be produced per year for the fixture to be economical?

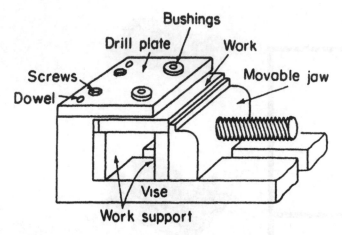

Figure 16.6 The use of a vise as a clamping device and as a jig for drilling holes. (Reprinted by permission of Prentice Hall, Upper Saddle River, NJ, from Ref. 3).

$$N = (4000 \times 0.60 + 2500 \times 0.25)/0.45(1 + 0.40)$$
$$= 4802 \text{ pc/yr}$$

This implies that at least 4802 pieces should be made per year for each year in the investment life of the fixture, that is, for 3 yr. If the total expected production over the 3-yr period is less than 14,406 pieces, the fixture should not be made. Expressions such as Eq. (16.3) are approximations; more accurate expressions can be developed for each company.

16.4 DIES FOR MANUFACTURING

A die is a tool containing a cavity that imparts a shape to a solid, molten, or powder metal or other material, such as wax or plastic, Thus, for permanent molding or die casting, the material is liquid metal; for investment casting, the molten wax is molded into the desired shape. In forging, wire drawing, and extrusion, the material is solid; in the powder metallurgy process, powder is formed into its shape. In plastic molding, the material is a semisolid as it is injected into the mold.

One of the design considerations in dies for permanent mold casting, die casting, and forging is that draft must be added to the design so the part can be removed from the die. *Draft* is a taper on the cavity walls so the part can be removed from the cavity after it is formed. A second concern is where to locate the parting line to form the two die halves so the part can be easily removed after forming. Thermal expansion of the die material must be considered as well as thermal contraction of the part during cooling in controlling part dimensions.

The cost of dies for manufacturing is difficult to determine, but one procedure (8) is:

$$DC \text{ (total)} = DC \text{ (material)} + DC \text{ (machining)} \quad (16.4)$$

$$DC \text{ (material)} = A + B \times A_c t^{0.4} \quad (16.5)$$

$$DC \text{ (machining one cavity)} = MR \text{ ($/hr)} \times M_t \quad (16.6)$$

where

A = equation constant
B = material cost constant
A_c = surface area
t = cavity thickness
MR = labor and overhead rate ($/hr)
M_t = machining time for cavity

If more than one cavity is made in the die, the machining cost would be for Y cavities:

$$DC \text{ (Y cavities)} = DC \text{ (1 cavity)} \times Y^{0.7} \quad (16.7)$$

Example Problem 16.2. A simple die casting die is to be made with four cavities. The material equation constant is $1000; the material cost constant is 0.024 when the surface and cavity thickness are in mm. The cavity surface area for each of the cavities is 2500 mm², and the cavity thickness is 40 mm. The machining rate is $600/hr; the machining time for a cavity is estimated to be 8 hr. What is the estimated die cost?

Solution. The machining cost per cavity would be:

DC (machining one cavity) = $600/hr × 8 hr = $4800

The machining cost for four cavities is:

DC (4 cavities) = $4800 × $4^{0.7}$ = $12,670

The die material cost would be:

$$\begin{aligned} DC \text{ (material)} &= 1000 + 0.024 \times 10,000 \times 40^{0.4} \\ &= 1000 + 0.024 \times 10,000 \times 4.37 \\ &= 1000 + 1050 \\ &= 2050 \end{aligned}$$

The total cost for the four-cavity die would be $14,720. The constants are dependent upon the complexity of the die, and dies with large complex surfaces would have a higher equation constant. The cost for forming dies and powder metallurgy dies would have different constants and exponents. The die costs for complex dies for aluminum engines can exceed $500,000.

16.5 PATTERNS FOR SAND CASTINGS

Patterns produce the image of the desired part in the molding material, which usually is sand. The design of the patterns is somewhat similar to that of dies, in that two surfaces must be generated, one for the cope half of the mold and one for the drag half of the mold. However, a pattern shape is like that of the part being produced, whereas a die has the *inverse* image of the part. The pattern produces a cavity in the molding material that has an image similar to the die image; the sand mold is the "die" for the liquid metal. Since patterns do not need to withstand the temperatures and/or stress levels of dies, their cost is usually lower that that for dies.

The materials used for the pattern depend upon the number of molds to be produced. If only a few castings are to be made, the pattern may be made of wood, urethane, or epoxy. If a moderate production level is anticipated, the pattern would be made of aluminum. For high production levels, iron and steel patterns are used, and the patterns are chrome plated to increase wear resistance.

The tolerances for castings tend to be larger than for other processes, because the metal is initially liquid and contraction usually occurs in the liquid state, during solidification, and during the solid state. The shrinkage during casting cooling is not uniform in all directions; thus it is difficult to predict shrinkage accurately.

The pattern costs would be similar to those for dies, but the pattern material used would have a major impact upon the total costs. The use of wood patterns would require less machining capabilities and the material cost would be lower than that of chrome-plated steels:

$$PC \text{ (total)} = PC \text{ (material)} + PC \text{ (machining)} \tag{16.8}$$

$$PC \text{ (material)} = A + BA_c t^{0.4} \tag{16.9}$$

$$PC \text{ (machining one cavity)} = MR \text{ ($/hr)} \times M_t \tag{16.10}$$

where

A = equation constant
B = material cost constant
A_c = surface area
t = die height
MR = labor and overhead rate ($/hr)
M_t = machining time for one cavity

If more than one cavity is made in the die, the machining cost for Y cavities would be:

$$\text{PC } (Y \text{ cavities}) = \text{PC } (1 \text{ cavity}) \times Y^{0.7} \qquad (16.11)$$

Example Problem 16.3. A pattern is to be made with four castings. The pattern is to be a chrome-plated steel pattern for high production. The material equation constant is $1500; the material cost constant is 0.035. The cavity planar area for each cavity is 3600 mm², and the casting height is 150 mm. The machining rate is $400/hr, and the machining time per cavity is estimated to be 6 hr. What is the estimated pattern cost?

Solution. The machining cost per casting pattern would be:

$$\text{PC } (\text{machining 1 casting cavity}) = 400 \times 6 = \$2400$$

The machining cost for four cavities would be:

$$\text{PC } (4) = 2400 \times 4^{0.7} = \$6334$$

The pattern material cost would be:

$$\text{PC } (\text{material}) = 1500 + 0.035 \times (4 \times 3600) \times 150^{0.4}$$
$$= 1500 + 3740$$
$$= 5240$$

The total pattern cost would be $11,574 for the pattern that would make four castings per mold.

16.6 TOOLS FOR SHEET METAL WORK

Major operations in sheet metal work include piercing, blanking, bending, forming, and drawing. Punch presses and dies are used for sheet metal work, and several sheet metal operations are often combined in a single- or multiple-station die. The punch is attached to the punch holder (moving part), and the die block is attached to the die holder (stationary part). Guide pins are attached to the die plate to allow the punch plate to activate up and down with respect to the die.

The cutting action between the edges of the punch and the die on sheet metal is a shearing process. The metal is subject to a shearing stress until it fractures. The clearance between the punch and the die allows the punch to push metal in the form of an embossed pad into the die opening. The cross-sectional area between the punch and die decreases, and eventually fracture (shearing) occurs. In piercing (hole cutting) the cut portion pushed into the die opening is scrap, whereas in blanking the cut portion is the workpiece. Figure 16.7 illustrates the shearing operation for a typical punch-and-die operation.

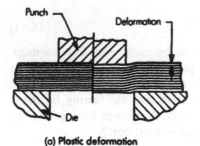

(a) Plastic deformation

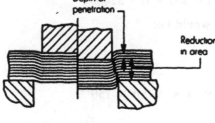

(b) Penetration

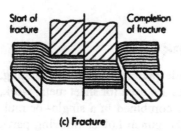

(c) Fracture

Figure 16.7 Shearing operation states of plastic deformation, penetration of cutting edge into the part, and fracture. (Reprinted with permission of the Society of Manufacturing Engineers, from *Tool and Manufacturing Engineering Handbook*, 4th ed., Volume 2. Copyright 1984.)

A progressive die is a multistation die. The dies are used where a single die would be too complex to do the job. In a progressive die, the sheet strip is moved through a series of dies, each performing a distinct operation. All dies are activated with each stroke of the press ram, and a finished part emerges from the last station with each stroke of the ram. An illustration of the workpiece and developed blank are shown in Fig. 16.8. There are four stations in the progressive die, two parts are produced with each stroke, and no carrier strip or skeleton was needed.

There are numerous tools for sheet metal working, and the costs for some of the different tools are given by Boothroyd et al. (8), for bending, piercing, and blanking. The approach to the tooling cost for sheet metal working is the same as that for dies for forming and die casting; that is, the tooling cost is the sum of the material cost and the machining cost to form the tool. The example for bending will be used to illustrate the procedure.

The total cost for the tool is the sum of the material cost and machining cost and can be represented as:

$$TB \text{ (total)} = TB \text{ (material)} + TB \text{ (machining)} \tag{16.12}$$

where

$$TB \text{ (total)} = \text{total tooling cost for bending}$$
$$TB \text{ (material)} = \text{material cost for tooling}$$
$$TB \text{ (machining)} = \text{machining cost to form tooling}$$

The expression (8) for the material cost is:

$$TB \text{ (material)} = A + B \times A_u \tag{16.13}$$

where

A = equation constant for tooling
B = material cost constant
A_u = usable area for tooling

The expressions for the machining cost for the tooling are:

$$TB \text{ (machining)} = C + (X1 + X2) \times MR \tag{16.14}$$

$$X1 = 0.68 \times L_b + 5.8N_b \tag{16.15}$$

$$X2 = (18 + 0.023L \times W) \times (0.9 + 0.02D) \tag{16.16}$$

where

C = machining constant for tooling
MR = machining rate (labor and overhead), $/hr
$X1$ = tool shape factor time, hr

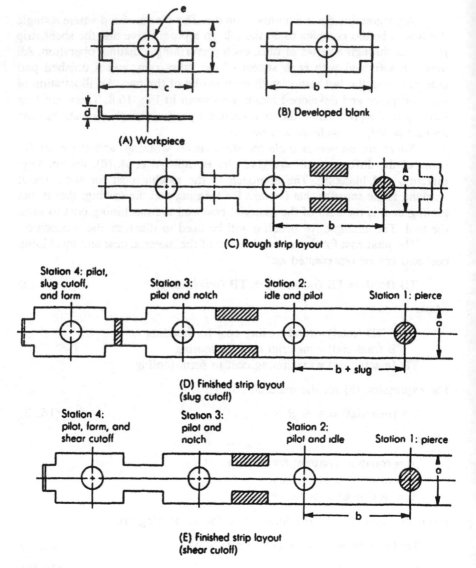

(A) Workpiece

(B) Developed blank

(C) Rough strip layout

Station 4: pilot, slug cutoff, and form
Station 3: pilot and notch
Station 2: idle and pilot
Station 1: pierce

(D) Finished strip layout (slug cutoff)

Station 4: pilot, form, and shear cutoff
Station 3: pilot and notch
Station 2: pilot and idle
Station 1: pierce

(E) Finished strip layout (shear cutoff)

Figure 16.8 A four-station progressive die with no corner strip. (Reprinted with permission of the Society of Manufacturing Engineers, from *Fundamentals of Tool Design*, 3rd ed. Copyright 1991).

$X2$ = part shape factor time, hr
L_b = length of bend, cm
N_b = number of different bends to be formed in die
L, W = length and width (respectively) of rectangle that surrounds part (cm)
D = final depth of bent part, cm (minimum value of 5.0 cm)

To illustrate the process, a example problem similar to that originally presented by Boothroyd (8) will be evaluated. The problem is somewhat different, for more data is needed than that included in the original example. Consider the shape in Fig. 16.9, to be formed in bending. The total rectangular sheet shape is 44 × 24 cm; the total number of bends is 5; and the bend length would be 3 × 10 cm + 2 × 23 cm, or 76 cm. The final height of the part is estimated to be 12 cm.

Using Eq. (16.16), the value of $X2$ is:

$X2 = (18 + 0.023 \times 44 \times 24) \times (0.9 + 0.02 \times 12)$
$= 48.2$ hr

Using Eq. (16.15), the value of $X1$ is:

$X1 = (0.68 \times 76 + 5.8 \times 5)$
$= 80.7$ hr

If the machining constant for bending is $780 and the machining rate is $40/hr, the machining cost using Eq. (16.14) would be:

TB (machining) = $780 + 40 \times (48.2 + 80.7)$
$= \$5936$

The die set for bending usually has a clearance of approximately 5 cm in all directions. Thus if the sheet dimensions are 44 cm × 24 cm, the die set dimensions would be 54 cm × 34 cm. The cost for the materials for the tooling, if the equation constant is $120 and the material cost constant is 0.36, then the material cost for tooling, using Eq. (16.13), would be:

TB (material) = $120 + 0.36 \times 54 \times 34$
$= \$120 + \$661$
$= \$781$

The total cost of the die set would be the sum of the costs, $5936 + $781 = $6717. Other expressions for piercing operations can be developed from the work of Boothroyd et al. (8). The tooling costs for sheet metal working tend to be lower that those for the hot forming processes such as die casting, permanent molding, and forging.

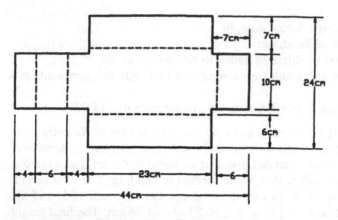

a) Block Sheet for Bending.

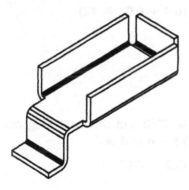

b) Shape after Bending.

Figure 16.9 Sheet and formed part for bending. (Adapted from Ref. 8, p. 380, and reprinted by courtesy of Marcel Dekker, Inc.)

16.7 TOOLS FOR INSPECTION AND GAGING

A manufactured product must be measured to ascertain that the product is within design tolerances and specifications. The key to quality control is to make certain the process is in control, and measurements of the product are done to evaluate the performance of the process. The equipment for inspection can be classified (9) into four different categories:

General-purpose measuring devices
Fixed gages
Visual reference gaging
Automatic gaging

16.7.1 General-Purpose Measuring Devices

General-purpose measuring devices can be further divided into nongraduated instruments, graduated instruments, and comparative instruments. Nongraduated instruments include calipers, dividers, telescoping gages, straightedges and squares, surface plates, and sine bars. Nongraduated instruments are used for comparing measurements rather than for making a direct measurement.

Graduated instruments include rules, vernier calipers, height gages, depth gages, micrometers, and protractors. Many of the instruments now include direct digital readouts, which make reading of the instrument easier.

Comparative instruments include mechanical indicating gages (frequently called *dial indicators*), pneumatic indicating gages, and electronic indicating gages. These instruments compare the workpiece being measured against a master used to calibrate the instrument. They measure the amount of deviation from the calibrated size, such as a gage block.

16.7.2 Fixed Functional Gages

Fixed functional gages are designed for a specific application rather than for obtaining a specific measurement. They are the reverse physical replica of the workpiece being measured and usually are designed to measure a single dimension. The gage may be used to indicate the normal condition or one of its limiting conditions.

One of the most common types of gages is the GO and NOT-GO (also called NO-GO) gages. The GO gage checks the part at its maximum material limit; that is, the part should fit inside the gage if one is measuring an external dimension, or the gage should fit inside the part if one is measuring an internal dimension such as a hole. The NOT-GO or NO-GO gage checks the minimum-material limit; that is, the part should not fit inside the gage if one is measuring an external dimension, or the gage should not fit inside the part if one is measuring an internal dimension. The major advantage of the GO and NOT-GO gages are the short time it takes for a check of the part dimensions.

There is a wide variety of GO and NOT-GO gages; the most common are: cylindrical plug gages, cylindrical ring gages, snap gages, taper gages, and thread gages. Snap gages are very common and can be used to check

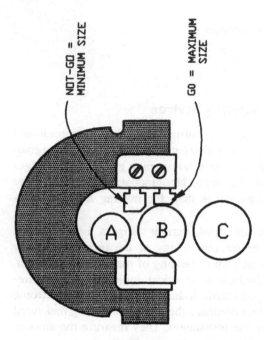

NOT-GO = MINIMUM SIZE

GO = MAXIMUM SIZE

A = TOO SMALL
B = CORRECT SIZE
C = TOO LARGE

Figure 16.10   Snap gage for measuring diameters of cylindrical shafts.

measurements of shafts very quickly. The gage illustrated (10) in Fig. 16.10
permits very quick measurement of the diameter; the part should be able to
pass through the GO gage limit but not through the NOT-GO limit.

Plug gages are precision-ground cylindrical gaging devices. These
gages are available in sizes ranging from 0.011 in. to 1.000 in. in increments
of 0.001 and 0.0005 in. in standard 2-in. lengths (4). Plug gages can be used
to check hole sizes, slot sizes, hole indicators, and the distance between
holes. The GO and NOT-GO plug gages are illustrated in Fig. 16.11; the
NOT-GO gage is usually colored red, and the GO gage is usually colored
green.

16.7.3   Visual Reference Gaging

Visual references consist of transparent materials that are dimensionally stable and can accurately show feature sizes and/or position. The typical visual
references include optical comparators, microscopes (toolmakers' micro-

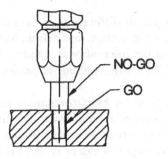

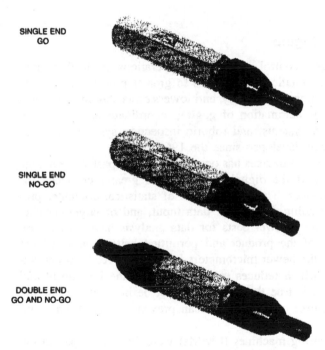

SINGLE END
GO

SINGLE END
NO-GO

DOUBLE END
GO AND NO-GO

Figure 16.11   GO and NOT-GO plug gages. (Courtesy of Carr Lane.)

scope), plastic overlay templates, optical flats (use of interference fringes), and laser inspection devices. The advantages of these devices are that frequently many dimensions can be checked simultaneously, entire surface profiles can be checked rather than only specific locations, and no gaging pressure or surface contact is required.

The use of optical flats can permit differences in measurement related to the wavelength of light emitted by materials, and helium has a wavelength of 23 μin. Although krypton-86 is the international standard for measuring length and has a wavelength of 23.8 μin., helium is preferred because it costs less and is easier to use.

Lasers differ from ordinary light by being extremely intense and coherent (coordinated in time and space) and having essentially the same wavelength. A number of different gases are used, such as carbon dioxide, helium-neon, and argon. Some of the advantages of lasers is that they can measure large distances (such as for surveying), can easily convert light variations/ interruptions into electrical signals (for optical scanning), and exhibit instrument stability that permits their use in harsh, industrial steel mills as well as in laboratories.

16.7.4 Automatic Gaging

The desire for automatic control of inspection operations has lead to many new developments. Automatic control leads to greater process control and improves productivity, improves quality, and lowers costs. Automatic gaging includes topics such as automation of gaging, coordinate measuring machines, machine vision systems, and robotic inspection systems. The last three systems have been developed since the 1960s.

Automation of gaging devices has occurred over the past 70 years, but the rapid development of the digital computer has greatly enhanced the application of such devices. The development of statistical computer programs, of electronic reading devices for data input, and of larger-capacity computer systems at much lower costs for data analysis have lead to improved measurements of the product and permitted better control of the process. For example, the newer micrometers have digital readings and automatic data logging, which reduces errors in reading the instrument and recording the values. It is true that quality cannot be inspected inside the product, but better control of the process can prevent defects from being produced.

Coordinate measuring machines (CMMs) were first developed in the early 1960s and have led to improved accuracy, since all measurements are taken from a common, geometrically fixed measuring system. The measuring of all significant features in one setup prevents errors due to setup changes. Coordinate measuring machines have reduced setup time because of only one setup rather than several; this also improves productivity and data logging for quality records and certification.

Machine vision systems began serious application in the early 1970s as computer and data imaging systems made significant improvements. The main inspection applications of machine visions systems are for gaging, verification, and flaw detection. Gaging is done to ensure that certain dimensions are acceptable. Verification involves determining that a product is complete (all parts of the product have been assembled) or in its proper orientation (for automatic assembly operations, the parts must be properly oriented). One of the problems with the early machine vision systems was that the image was two-dimensional; but the newer systems are able to consider three-dimensional images (for example, measuring a corner radius on an internal corner).

Robotic inspection systems allow flexible inspections, in that a variety of sensors can be mounted on the robot wrist or arm. The movement of the robotic arm reduces the need for precision fixturing. However, the typical inspection robot can inspect at a moderate level, such as 100 parts per hour, and are not well suited for a high-production system. The positioning repeatability of robots are from +0.025 mm to 0.20 mm, which would reduce the repeatability of the specific measuring system.

The computer revolution is improving the capabilities of automatic gaging systems, and rapid improvements in coordinate measuring systems, machine vision systems, and robotic inspection systems will occur. Not only will accuracy improve, but also the flexibility of the systems will increase.

16.8 SUMMARY

Tool design and engineering is an extremely broad field. Six major areas of tool design and engineering are machine cutting tools, jigs and fixtures, die design, pattern design, forming and cutting tools, and gages and measuring systems. It is such a broad area that tool engineers are specialists in only one or two of the topics presented. Tool engineering is one of the critical engineering areas that has been neglected, an area that requires specialization and is critical to the long-term manufacturing capabilities of a company.

Computers are having a major impact in automatic gaging, especially in the areas of coordinate measuring systems, machine vision systems, and robotic inspection systems. As computer power increases and computer costs are reduced, the capabilities of these flexible gaging systems will be further enhanced.

Tooling is frequently the dominant cost in many processes, such as die casting, forging, and plastic molding. Simplified tooling cost expressions were presented to indicate the two major cost items in tooling: the tool material cost and the machining costs to produce the tool. Jigs and fixtures usually must be evaluated on an economic basis, that is, whether the savings from the use of the fixture offset the investment costs for the jig or fixture.

16.9   EVALUATIVE QUESTIONS

1. What are the six most common types of tooling?

2. A milling steel cutter is 4 in. in diameter. What is the expected number of teeth for the cutter? (22)

3. What is a fixture? What is a jig?

4. What are the twelve directional movements, or degrees of freedom?

5. What devices are used for restricting workpiece movement?

6. It is estimated that 10,000 parts are going to be made per year for the next 3 years. The fixture cost rate is 60 percent, the salvage value at the end of the year is 40 percent of the investment, the total overhead rate is 50 percent, and the savings per piece are estimated to be $1.00. What is the maximum that should be spent for the fixture if the return rate is 25%?

7. What three design considerations are mentioned for making dies for forging?

8. A six-cavity die is to be made for diecasting. The material equation constant is $4000, the material cost constant is 0.040, the cavity surface area is 4000 mm^2, and the cavity thickness is 200 mm. The machining time ($300/hr) per cavity is 20 hr. What is the expected tooling cost for the six-cavity die?

9. A pattern is to be made with ten castings. The pattern is to be a chrome-plated steel pattern for high production. The material equation constant is $2000, and the material cost constant is 0.040. The cavity planar area for each cavity is 2500 mm^2, the casting has a depth of 75 mm, and the machining rate is $300/hr. What is the expected tooling cost for the ten-cavity pattern? The machine time for a single cavity is 6 hrs.

10. The material used for a casting pattern depends upon the production quantity. What materials are used for low production quantities, and what materials are used for high production quantities?

11. A carbide tool cutter is 6 in. in diameter. What is the expected number of carbide inserts (teeth) to be used for the cutter?

12. What are the four different categories or equipment for inspection and gaging?

13. What do the GO and NOT-GO gages measure with respect to material limit?

14. A bending die is to be made to form a Z-shaped part (two bends). The total area of the sheet is 10 cm × 20 cm, the number of bends is two, the

bends are both on the 10-cm side, and the final part height would be 10 cm. The machining constant for bending is $180, and the machining rate is $30/hr. The die set clearance is 5 cm, the material equation constant is $100, and the material cost constant is 0.36. Determine the total cost of the bending die set.

15. What are the four major systems of automatic gaging?

16. A cylindrical rod is held between centers on a lathe. How many degrees of freedom are restricted? What are the degrees of freedom that remain?

17. A progressive die is used to make a part. Why must the distance from one die station to the next die station be constant?

REFERENCES

1. Donaldson, C., LeCain, G., and Goold, V. C. Tool Design, 3rd ed., McGraw-Hill, New York, 1973, pp. 1–6, 573–577.
2. Fundamentals of Tool Design, 3rd ed., Society of Manufacturing Engineers, 1991, p. 755.
3. Pollack, Herman W. Tool Design, 2nd ed., Prentice Hall, Englewood Cliffs, NJ, 1988, p. 330.
4. Carr Lane Manufacturing Company. Jig and Fixture Handbook, 1st ed., St. Louis, MO, 1992, p. 430.
5. Jain, R. K., and Gupta, S. C. Production Technology, 12th ed., Khanna Publishers, Delhi (India), 1991, p. 1827.
6. Carpenter Technology Corporation. Guide to Machining Carpenter Stainless Steels and Specialty Metals, Carpenter Technology Corporation, Reading, PA, 1993, p. 155.
7. Kennametal Turning Products. Kennametal, Latrobe, PA, 1991, p. 404.
8. Boothroyd, G., Dewhurst, P., and Knight, W. Product Design for Manufacture and Assembly. Marcel Dekker, New York, 1994, pp. 339–352, 436–440.
9. Wick, Charles, and Veilleux, Raymond, eds. Tool and Manufacturing Engineers Handbook, Vol. 4: Quality Control and Assembly, 4th ed., Society of Manufacturing Engineers, Dearborn, MI, 1987, pp. 3-1–3-62.
10. Johnson, H. V. Manufacturing Processes, 2nd ed., Bennett Publishing, Peoria, IL, 1984, p. 223.

INTERNET SOURCES

Carr Lane home page—typical jigs, fixtures, and design sources: *http://www.carrlane.com/*

17
Failure Analysis and Prevention

17.1 INTRODUCTION

The failure of a part or component can lead to serious consequences involving loss of life, injury, property damage, or financial loss as well as loss of reputation and other "goodwill" attributes. Some of the more notable "failures" have included the *Titanic*, the "unsinkable ship" that sank when it hit an iceberg, and the *Challenger* space shuttle, which exploded during launch in 1986. There are usually two different views when a failure occurs: (1) the legal viewpoint, that is, ascertaining whose fault it is so those injured can obtain appropriate (financial) recourse, and (2) the engineering viewpoint, that is, ascertaining what went wrong and how it can be prevented from recurring. This is not much different from the criminal legal system: (1) The lawyers, jury, and judge try to determine if the accused is guilty and how much the guilty person should be punished (in terms of money and time), and (2) the police (like the engineers) try to solve the crime (find the source of failure) and to prevent it from recurring (remove the source). This chapter will focus primarily on the engineering viewpoint; but some knowledge of the legal viewpoint is necessary to prevent unnecessary financial losses.

17.1.1 Sinking of the *Titanic*

The *Titanic* (1) hit an iceberg at approximately 11:40 PM on April 14, 1912, and sank approximately 3 hours later. The *Titanic* was nearly 900 feet in length, weighed more than 46,000 tons, and was considered the state of the technology in shipbuilding. The ship was on its maiden voyage from Southampton, England, to New York City. There were more than 2000 passengers on board; more than 1500 died in the accident.

The ship was designed to be "unsinkable," in that the hull was built with separate compartments that could be sealed off from each other. If one of the compartments was punctured, that compartment could be sealed and the ship could continue sailing. In the worst case, if two or three compartments were flooded, it would take two to three days to sink and help should arrive. The problem was that in the accident, six compartments were flooded in the front of the ship, which caused the rapid sinking of the *Titanic* and high loss of life before help could arrive.

The ship was not found until 1985, and samples were not obtained until 1991 to determine what had happened. However, there were some indications of the cause of the failure from other ship sinkings by brittle fracture, such as the Liberty Ships during World War II. The *Titanic* was new and the sinking occurred rapidly, which indicated brittle fracture rather than ductile fracture. The Liberty Ships that sank were in cold water and had impact loading (such as a wave in a storm or hitting a pier at dock during rough water conditions). The Liberty Ship failures were analyzed, and the conclusion was that the steels became brittle at temperatures near freezing. Verification of the cause of the sinking of the *Titanic* could not be found until the ship could be located and the steel tested. Testing of the steel was performed using the Charpy impact test, and the steel was found to be brittle when tested near freezing conditions. Examination of the steel by metallography revealed that it had high levels of sulfur inclusions, which was not uncommon in the early 1900s. The high amount of sulfur inclusions, and sulfur in the metal, made the steel brittle at low temperatures. The transition temperature (when steels change from ductile to brittle behavior) has decreased from the 25–35°F range to lower than −40°F so this failure is unlikely today.

The analysis of the failure could have been performed earlier by getting samples from the steel manufacturer and testing them, but failure analysis techniques were in their infancy back then. Also, it was very convenient to blame everything on the iceberg rather than determine the actual cause of the failure. Before 1916, it was difficult to hold manufacturers liable, and lawyers did not pursue product liability issues as they do today. In the 1916 case of *MacPherson v. Buick Motor Co.* (2), the courts held that negligence could be maintained against a manufacturer. This case marked the beginning of modern product liability issues.

17.1.2 Space Shuttle *Challenger*

The explosion of the space shuttle *Challenger* on January 28, 1986, represents one of the greatest product failures, not only because of the loss of life, but because it was observed by so many people. The failure resulted in an intensive study (3,4) to determine the cause of the accident. The flight was to have lasted 144 hours and 34 minutes, but the flight lasted approximately 73 seconds after launch. The entire crew—Francis R. Scobee, Michael John Smith, Ellison S. Onizuka, Judith Arlene Resnik, Ronald Erwin McNair, S. Christa McAuliffe, and Gregory Bruce Jarvis—died in the tragedy. The failure caused a severe delay in manned space flight, with the next mission occurring 2.5 years after the explosion of the *Challenger*.

The tragedy was intense, because the nightmare took place in the sky and was visible for miles along the Atlantic coast. Screams of horror came from thousands of observers as the solid rocket boosters flew off in opposite directions. Families of the crew members were unable to believe what they were seeing. Two of the crew members, S. Christa McAuliffe and Gregory Bruce Jarvis, were payload specialists; that is, they were not career astronauts. Christa McAuliffe was a school teacher who was to conduct a series of classroom lessons and experiments from orbit. Mr. Jarvis was a satellite design engineer for Hughes Aircraft who was to perform a series of fluid dynamics experiments for satellite redesign.

I remember that day, for it was the day I was to take my first flight across the Atlantic. The accident happened a few hours before my flight, and I had second thoughts about going to Europe.

The Presidential Commission on the Space Shuttle *Challenger* Accident issued their report on June 6, 1986, four months after the formation of the commission on February 6, 1986. The commission (3) concluded that cause of the *Challenger* accident was the failure of the pressure seal in the aft field joint of the right solid rocket motor. The failure was due to a faulty design unacceptably sensitive to a number of factors (3): the effects of temperature, physical dimensions, the character of materials, the effects of reusability, processing, and the reaction of the joint to dynamic loading.

In addition to the technical component failure, there were serious flaws in the decision-making process. Two crucial areas of relevant concerns at Level 3 (project managers at Marshall and Johnson Center) were not communicated to Level 2 (preflight readiness review) or Level 1 (flight readiness review). These two areas were:

1. The objections to launch voiced by Morton Thiokol engineers about the detrimental effect of cold temperatures on the performance of the solid rocket motor joint seal.
2. The degree of concern of Thiokol and Marshall about the erosion of the joint seals in prior shuttle flights, notably those of January 1985 and April 1985.

The *Challenger* had flown nine times prior to its fatal last flight, whereas the *Columbia* had made seven trips, the *Discovery* six trips, and the *Atlantis* two trips. Thus, a freeze was made on shuttle travel until the cause of the *Challenger* accident was accurately determined, since the other shuttles would require the same type of launching system. Thus the accident resulted not only in the failure of one mission, but also in the cancellation of several other scheduled missions. The delays and cancellations were very costly, not only financially, but also in terms of the reputation and integrity of the space program.

On February 11, 1986, less than a week after the commission was formed, Dr. Richard Feynman, professor of physics at the California Institute of Technology and a 1965 Nobel Laureate in physics, performed a simple test that gave a strong indication of the problem. He put a sample of the seal material in ice water and detected that it had no resilience. The launch occurred on the morning after the temperature had been in the low 20s, and large amounts of ice had formed. The presidential commission did a very thorough investigation; the results were not only a better-designed shuttle launcher, but a better management system as well.

The *Challenger* accident illustrates that failures are usually not due to simple, single causes. One failure was the technical failure of the material; a second failure was the lack of communication of information regarding the engineers' concerns about the seals; and a third failure was the decision to launch the space shuttle under weather conditions that were marginal.

17.2 FAILURE TYPES AND MODES

There are several sources on failure analysis (5–8). Two of the basic points are to describe the different types of failure and the general modes of failure. There are three different classes of failure, based upon component/system performance, and these classes have been promoted by the military:

1. *The part is inoperable*: Examples would be a computer that doesn't work; an automobile engine that won't start; and an airplane missile launcher that will not launch any of its missiles. It does not mean that the part is destroyed, only that it is not operable.

2. *The part no longer performs its intended function satisfactorily*: Examples are a computer printer that keeps jamming the paper; an automobile engine that stalls when it is put into gear; and an airplane missile launcher that will launch some, but not all, of its missiles. This implies that the part may operate, but not on a regular basis or in the intended manner.

3. *Deterioration has made the part unreliable or unsafe for continual use*: Examples are a computer printer that has a loose wire that may short-circuit and cause electrical shock or ignition of the graphite ink; an automobile engine that stalls after operating at 65 mph on an interstate highway; a missile launcher that may activate

the missile but not launch it. In this class, the unreliability and danger of the system can lead to catastrophic accidents.

A set of similar categories can be noted in MIL-W-8604-A of 15 March 1982, which defines the three different classifications of criticality in aluminum alloy (plus a subclass):

Class A—critical application: A weldment is critical where a failure of any portion would cause loss of system, loss of major component, loss of control, unintentional release of critical stores, or endangerment of personnel.

Class A (SP): A subclass of Class A for applications such as pressure vessels, cryogenic, hypergolic, and vacuum systems, which require special (SP) considerations.

Class B—semicritical application: A weldment is semicritical when the failure would reduce the overall efficiency of the system but loss of system or endangerment of personnel would not be experienced.

Class C—noncritical application: A weldment is noncritical where failure would not affect the efficiency of the system or endanger personnel.

Another classification system for failures is by the amount of deformation and wear. Designers generally limit loads to prevent excess deformation and fracture, and these are two of the classes. The classification of defects by failure mode (7) leads to the following groups.

1. *Excessive elastic deformation*: Although the loading is below the yield point, the deformation is too high. This implies that the modulus of elasticity is too low; and since this is basically a material property, the wrong material was selected or used.

2. *Excessive plastic deformation*: This can occur when the loading exceeds the designer's predicted limit or the user uses the component in a nonintended manner.

3. *Fracture*: This is the most undesirable failure, for it frequently is catastrophic. When there is excessive elastic or plastic deformation, the part has usually not caused complete system failure. When fracture occurs, other parts are frequently also damaged.

4. *Corrosion and wear of parts resulting in loss of part geometry*: This can occur in pipelines used to transport coal slurries, since coal is an abrasive material and will tend to wear the pipes at bends and elbows. In this instance, the corrosion or wear of the part causes the failure of the part.

5. *Corrosion and wear of products that prevents the operation of other parts*: This frequently occurs in automobile cooling systems.

The cooling water causes corrosion in the radiator, and the products of corrosion are transmitted throughout the cooling system. The products of corrosion then cause wear of the water pump and cause it to fail. Note that if the corrosion of the radiator were prevented, the failure of the water pump could be prevented.

17.3 SOURCES OF FAILURE

There is a wide variety of sources of failure, including design aspects, material selection, material processing, fabrication, assembly, inspection, testing, storage and shipping, service condition, and maintenance. Often it is not a single source but a combination of sources interacting that can cause the failure. The primary sources of failure (6,7) are classified into the following eight groups: design deficiencies, material deficiencies, processing deficiencies, assembly errors, improper operation, neglect, service conditions, and management attitudes.

17.3.1 Design Deficiencies

1. *Failure to consider adequately the presence of stress raisers at critical points*: That is, notches, sharp fillet radii, corners and edges, keyways and the like can greatly increase local stresses.

2. *Inadequate knowledge of the service loads and environment*: The difference between operating at low, constant speeds at full load may not be as severe as cycling the speeds and loads. One of the problems Japanese automobiles initially had was that they were not prepared for the "rust belt" conditions of the Great Lakes and the Northeast coast of the United States, and their initial cars had severe corrosion problems.

3. *Difficulty of stress analysis in complex parts and loadings*: This has been reduced somewhat by the finite-element models, but the interactions of systems of parts is still difficult to predict. Also, a variety of conditions may exist: In a jet engine the temperatures are changing as well as the loading, and it is difficult to model mechanical stress, fluid stress (from the gas flow), and thermal stress effects simultaneously.

17.3.2 Material Deficiencies

1. *A poor match between service conditions and selection criteria*: A material may be required to have good wear resistance; but there is not a material property that is a good predictor of wear resistance. The material property commonly used to indicate wear resistance is hardness, but this is often a poor indicator.

2. *Inadequate material data*: For high-temperature applications, the material properties are usually considerably lower, but there is little high-temperature material data. For metals there is a limited amount of data, but for nonmetals the data is very difficult to obtain.

3. *Incomplete material specifications*: There is always an emphasis upon management to lower product costs. Sometimes materials are substituted to lower costs when the complete material criteria have not been specified. For example, a material may be used in a wear application and a hardness level specified; a second material may be substituted that meets the hardness level but will not meet the wear requirements. The hardness level was an adequate wear specification for the first material but not for the second material.

17.3.3 Processing Deficiencies

Materials can be improperly processed and have cracks or other processing defects that can lead to failure. Inclusions in casting, laps and seams in forgings, porosity and inclusions in weldments, and residual stresses from machining are some of the defects that can occur in material processing operations that have been sources of failures.

17.3.4 Assembly Errors

Bolts that have not been properly tightened can work loose during operation and cause failure. Misalignment of shafts, gears, bearings, and seals are frequently a contributing factor to service failures. Parts may have tolerances that are too loose, which leads to mismatch in assembly.

17.3.5 Improper Operation

Operation of equipment without using proper conditions for startup or shutdown can lead to failure. For example, many instructors shut off a slide projector without leaving the fan on to cool the bulb properly; if not cooled properly, the bulb fails.

17.3.6 Neglect (Inadequate Maintenance and Repair)

Automobile manufacturers give maintenance service schedules, yet many consumers neglect to have their automobiles serviced on a regular basis. Oil needs to be changed and the oil filter replaced, because the oil becomes dirty and the dirt can cause accelerated engine wear. Steel tools left in corrosive environments is another instance of neglect; this usually requires cleaning and resharpening of the cutting edges.

17.3.7 Service Conditions

1. In many instances, products are used in conditions beyond those allowed in the design. For example, an extension bar may be added to a wrench to increase the torque, but the wrench may fail because it was designed to operate at its actual length rather than at an extended length.

2. General deterioration over time. A hammer will increase in hardness and have higher residual stresses due to work hardening when it hits an object. Over a long period of use (usually years), stresses will build up until eventually a piece will fracture off. The paint on an automobile will eventually chip due to surface impact with road dirt, bugs, and other objects; this is expected over time.

17.3.8 Management Attitudes

1. *Excess focus on cost*: Management will often want to substitute materials that are cheaper without ensuring that the material meets the design requirements rather than the material specifications.

2. *Excess focus on production*: The pressure to get the product out the door or to complete the job often leads to hasty decisions and disastrous results. One company substituted a casing pipe of 80,000 psi for a welding line pipe X-42 (which meant that it had a tensile strength of 42,000 psi). However, they ignored the fact the casing pipe is not to be welded and they required the fabricator to weld the pipe. When the part failed in operation a year later, they tried to claim that the fabricator welded the pipe improperly and ignored the fact that they supplied the material and told him to weld it or lose the business. The problem was that they wanted "to get the job done" and not wait for the proper materials.

Although most failures fall into the first seven categories, even then management is involved, since they are responsible for the execution of product design, product development, product manufacture, employee training, sales and marketing, service, and operation. The total quality management system emphasizes that most failures are the responsibility of management, for they have failed to supply the materials, training, equipment, and other items needed to do the job properly.

17.4 FAILURE ANALYSIS PROCEDURE

The purpose of failure analysis is to determine the cause of a failure; it is not to determine who or what is to blame for the failure. A failure is often the result of a primary cause and one or more contributing factors. For example, in the *Challenger* space shuttle failure, the failure was attributed to the failure of a pressure seal. The failure was due to a faulty design unacceptably sensitive to the effects of temperature, physical dimensions, material characteristics, reusability, processing, and dynamic loading. It was not a single design problem, but a combination of factors not properly considered.

There are usually five major stages (7) in the failure analysis procedure, but they are presented differently by the various authorities on failure analysis. Each of the stages will be presented in outline form, giving some of the specific items in the particular stage. These stages are: field inspection, background history, examination techniques, nondestructive testing, and mechanical testing.

17.4.1 Field Inspection of the Failure (Preliminary Examination)

A. Timing: The investigation should be performed as soon as possible after the failure occurs. Whenever their is an airplane crash, an investigative team is sent to the site by the FAA to examine the crash site. Unfortunately, in automobile accidents, the investigation does not usually start until after the lawyers start the legal proceedings, and the only information about the accident site is in the police report.

B. Documentation During the Field Inspection: The information must be recorded and documented as soon as possible, because deterioration of the components will occur due to weather, dust, oxidation, etc.

   1. Photographs (and videos)

      a. Take photographs of the area to indicate what may have influenced the failure.

      b. Take photographs of the failure. First take the photographs at a distance to get an overall view, and then move closer to focus on the failure site for more detail.

      c. Photograph the failure at different angles. The replay of sports events on television have indicated the importance of the angle of the view; this is also true in failure analysis.

      d. Make careful notes with respect to the photographs/video to document the site at a later date; court trials are usually at least 2 years after a failure has occurred.

   2. Statements of participants, observers

      a. Statements from people in the area should be taken as to what happened, the conditions immediately prior to the failure, as well as typical conditions. The emphasis is to focus on any irregular conditions or events prior to the failure.

C. Information: The information items to be obtained according to Dieter (7) are as follows:

    1. The pieces should be located relative to each other in an attempt to visualize how the part failed. However, the surfaces *should not be fit back together*; this will interfere with the examination of the materials' surfaces in determining the path of the fracture or in determining surface contaminants. In addition, a similar assembly that has not failed should be obtained for comparison purposes and to understand how the parts originally fit together.

    2. An attempt should be made to identify the origin of the failure, especially in fracture and fatigue failures. The origin of the failure will need to be determined; and the earlier it is determined, the sooner the cause can be determined. The direction of crack propagation and the sequence of failure should be obtained from chevron marks or other indicators of crack propagation.

    3. The orientation and magnitude of the stresses at the time of failure should be obtained from the operating data at the time of failure. Also, a study of the forces by modeling may be done later.

    4. The presence of any obvious material defects, stress concentrators, oxidation products, temper colors, melting, corrosion products, pitting, or other potential failure contributors should be located, photographed, and described for further detailed examination.

    5. The presence of any secondary damage not related to the main failure, such as tool marks, dents, and surface alterations, should be located, photographed, and described in detail for further examination.

D. Preservation of the Fracture Surface:

    1. The fracture surface must be preserved in its original failed condition. *Do not attempt to fit the parts back together*, because you will destroy the fractured surface.

    2. *Do not touch the fractured surface*, for you will contaminate it.

    3. Avoid corrosion of the fractured surface; dry the fractured surface with dry compressed air and place in a desiccator to avoid corrosion. Otherwise it will be difficult to determine whether corrosion contributed to the failure or occurred after the fracture.

    4. *Do not take samples from the fractured surface* except as a last resort. Take samples below or near the surface for composition analysis, but *do not destroy the sample* during the analysis. The defendants must have access to the sample; were it destroyed, it could not be admitted as evidence. Also, once destroyed it can no longer be available for analysis.

17.4.2   Background History

A complete background history of the failed component must be developed, because similar failures may have occurred. This information should be obtained as soon as possible and would be useful at the field evaluation, although usually this is not practical. Some of the items that should be included in the background follow.

A. Item name, identification numbers, owner, user, manufacturer or fabricator, site of manufacture/fabrication, date of manufacture/fabrication
B. Design standards and calculations used; drawings and modifications to drawings during manufacture or fabrication
C. Manufacturing and fabrication methods; codes and standard used; inspection standards and techniques applied; repair and inspection done during manufacture
D. Documentation on materials used; quality assurance techniques use, inspection standards and techniques applied
E. Service history; inspection of operating logs and records; services performed and scheduling of service
F. Discussion with operating personnel and witnesses to determine any unusual conditions or events prior to failure
G. Date, time, temperature, and environmental conditions at the time of failure; conditions prior to failure; and any extreme conditions over time
H. Operation conditions: temperature and pressures; static and dynamic loading conditions; vibrations and cyclic loading; corrosive and erosive conditions

17.4.3   Examination Techniques for Failure Analysis

There has been a wide variety of techniques to examine the fracture surfaces of materials. In addition to the evaluation of the fractured surfaces, the materials must be examined to determine whether they are the correct materials, whether the materials have been processed correctly, and whether the operating conditions were within the designed range.

A. Visual Examination: Observe the surface of the fracture and the path of cracks; look for signs of abuse, corrosion, or environmental effects; and make a general assessment of the basic design. This stage often determines what specific or detailed evaluations should be performed. A visual examination includes using magnifying glasses with powers in the 5×–20× range.
B. Specimen Selection: Select the specimens to indicate what one is trying to find, verify, or explain. This can be performed to verify the potential cause or to exclude potential causes.

C. Macroscopic Examination (Usually ≤100×): The purposes of macroscopic examination are to observe gross features of the fracture; for the presence or absence of cracks; for the presence of any gross defects; for the presence of corrosion or oxidation products.

D. Microscopic Examination

   1. The purpose of microscopic examination is to determine microstructure features, such as grain size, size and distribution of second-phase particles, the size of inclusions, the presence of corrosion products, and the presence of any undesirable microconstituents. When using microscopic examination, large areas cannot be covered easily, so the macroscopic techniques are used to indicate which areas should be examined in detail by the microscopic techniques.

   2. There are various types of equipment used. Some of the more common types are:

       a. Metallurgical microscope (50–500×, some to 1000×)

       b. Scanning electron microscope, or SEM (1000×–40,000×)

       c. Transmission electron microscope, or TEM (up to 10^6×)

E. Chemical Analysis: Chemical analysis is often required to determine what the impurities are or if the material is of the appropriate composition. Some of the techniques used to determine the chemical analysis for samples follow.

   1. Energy dispersive units attached to the SEM are used to determine the composition of impurities for embrittlement, stress corrosion, etc. (This is also referred to the *microprobe*, or electron-beam microprobe, *analyzer*.)

   2. Wet chemistry is the traditional method for evaluating the chemistry of metal samples.

   There are several other methods for determining the chemical analyses of materials, such as auger electron spectroscopy, secondary ion mass spectroscopy, x-ray photoelectron spectroscopy, and the laser microprobe. These methods are more specialized, and other references should be consulted for details.

F. Protection and Preservation of Fracture Surfaces: The surfaces must be protected to prevent the corrosion or oxidation that would occur over time. Since corrosion and oxidation occur rapidly, it is essential that the parts be documented in their original condition as soon as possible.

17.4.4 Nondestructive Testing (NDT) Techniques

Nondestructive testing is done to detect surface cracks and discontinuities and internal discontinuities. Although NDT techniques are used for failure examination, they are more commonly used to test products to ensure the quality of the component before usage. The six basic NDT methods follow.

A. Visual Examination: The visual examination in NDT examines for surface defects (*nonconformities* for the legal community), such as cracks, chips, indentations, and scratches. It also includes metrology to verify the size and tolerances are attained.

B. Magnetic Particle: This technique is used on ferromagnetic materials to find surface defects. Small magnetic particles are applied to the surface; when a magnetic field is applied, the magnetic particles will gather at the defect. The defect causes a leakage in the magnetic field, and the particles are attracted to the field leakage.

C. Liquid Penetrant: This technique uses a liquid to penetrate into surface defects. The surface liquid is then removed, and a surface coating is applied that will absorb the penetrant from the surface defects and indicate their location.

D. Eddy Current Inspection: This technique can be used on electrically conductive materials. It is an electromagnetic-induction technique. Changes in the field imply discontinuities on the surface or slightly below the surface. The technique can be highly automated for rapid inspection of items such as pipe surfaces.

E. Ultrasonic Techniques: Ultrasonics is one of the two techniques used to evaluate internal as well as surface defects. It does require high operator expertise in the interpretation of the signal responses. It is usually the less costly method for internal defects and is easier to use than radiography.

F. Radiography (X-Ray) Techniques: The x-ray technique is the other main method for evaluating internal defects. It has the advantages that the defect response is easier to explain and that a permanent record is available. Many critical government parts require complete x-ray testing; in those cases, it may increase the part cost by 100 percent. The disadvantage of the x-ray technique is that the testing must be performed in a lead-shielded room, so it is not an easily portable technique. Recent advances are making the technique more portable. But for extremely large parts, it is very difficult to apply this technique.

There are numerous other nondestructive tests being applied, such as holography, acoustic-emission inspection, microwave inspection, thermographic inspection, and leak testing. Some of these tests, such as leak testing and holography, tend to be used for inspection purposes rather than for failure analysis.

One of the interesting policies sometimes used in NDT inspection is the retirement for cause. Parts are returned for service on a periodic basis until a defect is detected by an NDT procedure. The return-to-service period must be less than the time for a defect to progress from the nondetectable state to the failed state. This can be used for parts expected to have a fatigue failure. However, in the times of cost cutting, inspections may be delayed and catastrophic failure may occur.

17.4.5 Mechanical Testing and Analysis Techniques

Mechanical testing is performed when sufficient material from the sample is available to make samples, otherwise, samples are made from similar components. Since mechanical testing is destructive, there are limits to the amount of testing. This must be planned carefully to avoid wasting the limited amount of material.

A. Hardness Testing: Hardness testing is a relatively simple test that involves a minimum amount of material surface. Thus it is frequently utilized, since it causes minimal damage to the part. It can be used to estimate the tensile strength of the material, to indicate the heat treatment, or to indicate if work hardening has occurred. There are numerous types of hardness tests; some of the more common tests are the Rockwell, Brinell, Scleroscope, Vickers, and Knoop hardness tests.

B. Tensile Testing: Tensile testing is the standard test for determining the tensile properties of materials; thus, tensile data can be used to determine the yield strength, tensile strength, fracture strength, modulus of elasticity, elongation, reduction of area, and tensile instability. Tensile testing is done to determine if there are any irregularities in the material properties of the failed component in comparison with the properties used in the design. Both engineering and true stresses should be determined and evaluated. However, this test is used more to evaluate materials before design rather than during failure analysis. This is a destructive test, so it is usually done on test samples to avoid destroying the product to make test specimens.

C. Impact Testing: Impact testing is a test used to determine the behavior of the material to rapid rates of loading, and it measures the amount of energy absorbed in the fracture. The two tests commonly used to measure impact energy are the Izod and the Charpy tests. Specimens that absorb large amounts of energy are generally classified as ductile materials, whereas specimens that absorb small amounts of energy are classified as brittle. Some materials, including steel, will become brittle at low temperatures. The temperature at which the material changes from ductile to brittle is called the *transition temperature*. The seal material on the *Challenger* space shuttle did undergo a similar change in resilience, and this is one of the causes of the failure. This is a destructive test.

D. *Fracture Analysis*: Fracture analysis is used to determine the cause of a failure with respect to the fracture surface. This process utilizes various techniques to examine the fractured surface to predict the loading and environmental conditions at the time of fracture. Some of the more common types of fractures observed follow.

1. *Ductile fracture*: Some amounts of elongation and reduction of area have occurred, and ductile fracture occurs over time. The crack growth is slow, and plastic deformation is noticeable.

2. *Brittle fracture*: Brittle fracture has rapid crack growth, and little, if any, plastic deformation occurs. Little energy is absorbed during the fracture. In metals, there are two types of brittle fracture: (1) transgranular cleavage, where the fracture is along the grain boundaries; and (2) intergranular cleavage, where the fracture goes through the grains.

3. *Fatigue*: Fatigue fractures begin as minute cracks and progress under the action of fluctuating stresses, which are common in rotating shafts.

4. *Creep and stress rupture*: These are failures that occur at elevated temperatures via the mechanism of creep. At high temperatures, designs are governed by the amount of creep. That is, a certain amount of plastic deformation is expected over the life of the structure. The life is governed by the total amount of strain that will occur before rupture or fracture.

5. *Stress corrosion*: Stress corrosion requires both tensile stress and a corrosive environment. The tensile stress accelerates the corrosion, the crack tip increases the stress, and the crack propagates through the material. The process can be either transgranular or intergranular brittle fracture.

6. *Hydrogen embrittlement*: In metals, and in particular in steel, hydrogen dissolved during melting or welding will try to come out of solution as the metal cools. If it is present, it can form hydrides that make the metal brittle. In welding high-strength steels, it is essential to keep hydrogen from the weld pool to avoid hydrogen embrittlement.

E. *Testing under simulated service conditions*: This is done whenever possible, but it is costly. For example, automobile bumper collisions at 5 mph involve crashing a car moving at 5 mph and hitting a barrier. Today, mathematical models are being constructed to simulate the service conditions and predict the results.

One problem with mechanical testing is that the test results indicate the material property, but the type of loading that the component experiences is frequently more complex in actual conditions. For example, the hardness test is a uniaxial compression test and the tensile test is a uniaxial tension test, whereas stresses are usually triaxial. The part geometry also has an effect with respect to how the stresses are distributed. New tests and new properties will be developed to evaluate better the materials and designs of the future.

17.5   PRODUCT LIABILITY ISSUES

Product liability issues have greatly benefited the legal profession, but consumers and manufacturers have had the balance of justice favor first one side and then the other. Initially manufacturers had the protection of the law with the 1842 case of *Winterbottom vs Wright* (2). In that case, an injured mail coach driver could not maintain an action against the manufacturer of the defective mail coach. The balance of justice changed with the 1916 case of *MacPherson vs Buick Motor Co.* In this case, negligence could be maintained against the manufacturer of an automobile with a defectively made wheel that broke and caused injury to the plaintiff. Recent moves in the U.S. Congress have been to restrict liability claims to protect manufacturers because of so-called unreasonable awards by juries.

17.5.1   Product Liability Cases

There are two basic cases of product liability: strict liability and negligence.

1. *Strict liability*: In the case of strict liability, proof must be shown in the following three areas:
   a.   The product was defective and unreasonably dangerous.
   b.   The defect existed at the time the product left the defendant (which could be any or all of the producer, fabricator, supplier, or seller).
   c.   The defect caused harm.

The key item is usually to show that the defect existed at the time the product left the defendant, and that it was not created during use or elsewhere. Also, if no harm occurs, one cannot be found liable for damages.

2. *Negligence*: There are three different types of negligence, of which only one needs to be considered; the three types are:
   a.   Failure to warn of danger (most common, about 60 percent of cases)
   b.   A concealed danger in the product
   c.   A design that was inadequate or failed to meet standards

The typical case is the failure to warn. If you purchase a hammer at a store, you will notice a label warning that you should wear safety glasses when using the hammer (to prevent chips from entering the eye). This label warns of the potential danger of a chip's occurring while using the hammer, and this is sufficient to prevent the manufacturer from being held liable. If, on the other hand, while one were raising the hammer to hit something the hammer head came off and hit someone, there is a potential liability case because the warning does not cover that event.

17.5.2 Design and Management Aspects to Prevent Product Liability Cases

Product liability cases are very expensive, time-consuming, and annoying for manufacturers. It is not just the legal defense costs and awards; it is also the stress created and the time consumed for designers and manufacturers that is costly. Thus it is important that measures be taken to prevent products from being manufactured that can lead to injury or cause damage. Some of the considerations that should be taken in the design stage (7,9,10) follow.

1. Get legal advice of the product liability aspects before issuing new products or designs. Incorporate legal developments in product liability into the design decision process. All designs should be evaluated for the risks of potential product liability problems. All potential hazards should be evaluated, with estimates of the potential for occurrence and the seriousness (cost included) of the occurrence.
2. Make certain that there is strict adherence to all industry and government standards. Conformance to standards does not protect the manufacturer from liability, but it indicates that an attempt was made to protect the consumer.
3. Thoroughly test the product to identify potential unsafe uses and how a part will fail. Products fail, but it is essential that when failure occurs, there is no danger.
4. Use quality control techniques to ensure that the product meets design specifications. A successful quality control system should greatly reduce the number of defective products reaching the consumer.
5. User instruction manuals should be an integral part of the product design process. Warnings should be presented for all potential hazards of improper operation. Warning labels on the equipment should indicate operating dangers and appropriate personal safety protection equipment.

6. Documentation of the design, testing, quality assurance, and instruction manuals and warnings is very important.
7. The effect of any design changes should be evaluated, not only with respect to improved functional performance, but also with respect to any new potential liability.
8. Engineered products often consist of many components. The changes in one component may effect the performance of a second component and its failure potential. Thus the total system failure must be evaluated, and some modes of failure may be preferred over others.
9. Design for a safe system failure whenever possible. The use of fuses/fuse boxes in homes have greatly reduced fires caused by electrical shorts.

The management aspects of product liability are crucial to creating the proper attitude to the prevention of product failure and product liability problems. As the saying goes, it is not what you say, but what you do. Everyone wants to avoid product liability, but actions must be taken. Some actions (7) follow.

1. The formation of a committee to verify that the corporation has an effective product liability loss control program. This would ensure that the designers are aware of the latest industrial and government standards and that an appropriate quality assurance program is in place.
2. The company must have adequate insurance protection and a product recall procedure if a defective product is detected. If a product has a problem, get the product off the market and have the problem corrected.
3. The company should set up a reporting system to detect any unsuspected product defects, quality deficiencies, or product hazards as soon as the product reaches the marketplace. This system should also detect minor complaints, for they may indicate potential abuses.
4. Management must develop a positive attitude to product liability prevention. It is not only good public relations; it is also cost effective and productive. The interference caused by a product liability suit adversely effects production, design, marketing, and sales.
5. Management must promote product liability prevention by providing training to designers to avoid product liability design problems. Use outside consultants to evaluate products and designs for product liability issues.

Management must realize that most product liability cases are the result of poor management decisions and systems. A defect may be produced because of a poor design or an error in manufacturing, but management should have a system to detect the poor design or manufacturing error. In many instances, management is in a rush to get the production out the door, so it sacrifices product quality. The need to get the product to market rapidly often reduces the product testing period. Management must then develop accelerated testing procedures that can accurately predict the long-term environmental effects of corrosion and oxidation. New computer models are now simulating product behavior during development to predict potential design problems.

17.6 SUMMARY

Failure analysis is essential to prevent product failures from occurring via the detection of causes and potential causes of failure and their elimination. Many serious failures with fatalities have occurred, two of which are the sinking of the *Titanic* and the explosion of the *Challenger* space shuttle. Failure analysis modes can be classified in several different ways: with respect to the user of the product (customer), with respect to the manner in which the failure occurred via wear, deformation, or fracture; and with respect to the source of failure, such as the designer, the material, or improper processing. The three classes of product failure with respect to the customer are: the part is inoperable, the part doesn't perform satisfactorily, and the part is unreliable or unsafe.

The classification of failure with respect to the manner in which the failure occurred is an attempt to determine "how" the product failed. The classification of failure with respect to the source of failure is an attempt to determine "why" the product failed. The primary purpose of answering the "why" should be to prevent future failures; but it is also the primary purpose of assessing blame in product liability cases.

Although product liability legal ramifications cause serious management problems, product liability concerns have led to safer consumer products. Parts will eventually fail, but product liability issues have assisted in getting unsafe products removed from the marketplace and in retiring products at the end of a useful life before failure occurs. Although engineering design and product manufacture are part of the problem in product liability, management decisions with respect to reducing costs, increasing production, and reducing product testing and development to reduce lead times are usually the indirect primary cause of most failures.

17.7 EVALUATIVE QUESTIONS

1. What were the primary causes of failure in the *Titanic* and *Challenger* space craft accidents?

2. What are the three different classes of failure based upon component performance?

3. What are the three different types of classifications of fusion welds according to criticality?

4. What are the five different types of failure mode?

5. What are the eight primary sources of failure?

6. What are the key stages in the failure analysis procedure?

7. Why is it important not to touch the fracture surface or to try to put the fractured pieces back together?

8. What are the six basic nondestructive testing techniques?

9. What are the two basic cases of product liability?

10. What are the three areas in which proof must be shown in a strict liability case?

11. What considerations in the design stage of a product should be followed to prevent product liability actions?

12. If a product is defective, what must be done?

17.8 RESEARCH QUESTIONS

1. Review a failure analysis case, and prepare a three-page summary of the failure, failure analysis methods used, cause of the failure, and liability incurred.

2. Prepare a three-page report on one of the newer NDT methods.

REFERENCES

1. Gannon, Robert. "The Titanic's Final Secret," Reader's Digest, August 1995, Pleasantville, NY, pp. 155–160.
2. Noel, D. W., and Phillips, J. J. Product Liability, West, St. Paul, MN, 1974, pp. 1–12.
3. Rogers, William P. Report of the Presidential Commission on the Space Shuttle Challenger Accident, June 6, 1986.

4. Lewis, Richard L. Challenger—The Final Voyage, Columbia University Press, New York, 1968.

5. American Society for Metals, Source Book in Failure Analysis, American Society for Metals, Metals Park, OH, 1974.

6. Metals Engineering Institute. Principles of Failure Analysis, Metals Park, OH, 1980.

7. Dieter, George E. Engineering Design—A Materials and Processing Approach, 2nd ed., McGraw-Hill, New York, 1991, pp. 589–611.

8. Boyer, Howard E., and Gall, Timothy L. "Failure Analysis," Metals Handbook—Desk Edition, American Society for Metals, Metals Park, OH, 1984, pp. 32-1–32-32.

9. Weinstein, A. S., Twerski, A. D., Piehler, H. R., and Donaher, W. A. Product Liability and the Reasonably Safe Product, Wiley, New York, 1978, pp. 136–145.

10. Colangelo, V. J., and Thornton, P. A. Engineering Aspects of Product Liability, American Society for Metals, Metals Park, OH, 1981, pp. 59–73.

INTERNET SOURCES

Challenger explosion, by Paula Jarvis: *http://www.stlcop.edu/~johnpais/challenger/paula.html*

Failure analysis basics: *http://www.stresseng.com/failure.htm*

Nondestructive testing: *http://www.golden.net/~protech/ndtestin.html*

Plastic failure analysis: *http://www.eastman.com/ppbo/design/ftable.shtml*

Product liability and legal reform: *http://www.bus.orst.edu/faculty/nielson/industry/prodliab/prodliar.htm*

Titanic failure analysis: *http://www.tms.org/pubs/journals/JOM/9801/Felkins-9801.html*

Typical nondestructive testing methods, New Zealand NDT Society: *http://www.winzurf.co.nz/ndta/ndtmeth0.htm*

Index